There's Something in the Air

Clemson University Aviation and Space Heritage

Otha H. Vaughan Jr.

This book is dedicated to my parents, who provided me with an education that allowed me to succeed in my chosen profession as an Engineer and Research Scientist.

It is also dedicated to my wife Betty, our daughters, and our grandchildren.

Something In These Hills

Times when so many things seem to be coming unglued are disquieting times. These are disquieting times.

It always intrigues me how nearly any specific condition of nearly any specific time can find some application in a book that, essentially, was handed down to us by word of mouth century after century.

I believe it says somewhere, "I will lift up mine eyes unto the hills from whence cometh my help."

My thoughts are wandering through these upper South Carolina hills that shelter the University that forms a common bond for many thousands of people who have studied here, or taught here or worked here.

There's something in these hills that has touched every one of them, something that has rubbed off on them in varying degrees, something that has built within the breasts of all Clemson men and women an enduring spark akin to an eternal pride.

There's something in these hills. It was here when a handful of fledgling faculty members greeted a relatively small band of 446 students more than 80 years ago. That was shortly after convict labor had completed an administration building and clock tower that still dominate these Blue Ridge foothills with a timelessness and serenity that impart inspiration and strength anew each time they are looked upon.

There's something in these hills that has endeared itself to an endless procession of administrators, teachers, students, secretaries and workmen. Hundreds of names pass through my consciousness, names of people who gave selflessly of themselves to build the institution nestled here and who at one and the same time mined the priceless something the hills contain and returned to them still more of it.

I have my names and I see once more the faces and feel again the beloved personalities of those that go with them. If you will but close your eyes and drift awhile, you, too will recall the names and faces and personalities who meant the most to you while the privilege of being among them was yours.

There's something in these hills and from them we have drawn the power to transcend the stresses and strains that tug away to make things come unglued in these disquieting times, the power to cut through such modern concepts - and such modern facts - as generation gaps, communication gaps and idealogical gaps.

Where is the generation gap when an alumus who spent four years in these hills before the turn of the century says, "Next to my church and my home, I love Clemson University beyond all other institutions this side of Heaven": and when a graduate-to-be says, "Excepting only my parents, Clemson has meant more to me and done more for me than anything that has touched my life"?

There's something in these hills that has bound together a man over ninety and a boy of under twenty, something that has given them common ground on which to stand and a start toward bridging, and eliminating, any gap or any stresses or any strain that might try to make unglued whatever they seek for themselves as they move out of these hills into the mountains, the plains, the oceans, the forests, the skies and the storms of life.

We have all drawn from these hills something to suggest to youth that those over thirty can be trusted and to indicate to those over thirty that the qualities of youth are as sound today as they ever were

There is something in these hills that brings together and binds together and holds together men and women of all persuasions, of all heights, sizes, weights and cultural backgrounds - something that cuts across every difference, spans every gap, penetrates every wall - something that makes a man or a woman stand taller, feel better and say with high pride to all within earshot, "I went to Clemson."

There is something in these hills that you and I can't define and others can't understand. A wave of warmth always surges through me when "outsiders" say, "I don't know what it is about you Clemson people, but your undying love for Clemson is admired by everyone I know."

There's something in these hills and I suspect that's what it is - the ability of an institution through the unending dedication and greatness of its people - its administration, its faculty, its staff, its students and its alumni - to impart to all it touches a respect, and admiration, an affection that stands firm in disquieting times when things around it give impressions of coming unglued.

Yes, there's something in these hills where the Blue Ridge yawns its greatness.

Joe Sherman '34
Executive Director
Clemson Alumni Association
1956-1977

PREFACE

This manuscript came about as my desire to research and compile the early Clemson Aero Club history and Clemson's Aviation and Space Heritage. I will discuss how aviation got started at Clemson A & M College in 1927 and how it has continued till the present, about how I became a General Aviation pilot, and other details that I could remember. Also the input from other people that knew about how aviation came to Oconee County and to Clemson has been very helpful. Since I was my instructor's first student to solo from the airport in Seneca, SC, I have included some general details about his biography, about myself in the early days of learning to be a General Aviation pilot at the Seneca Airport, and how the first airport came into being near the campus of Clemson A & M College in 1946. As I began researching the Aero Club's history it became clear to me that I would miss many details about the club and it's operation over the past years but I tried to the best of my ability to be as correct as possible. Ron Shelton of the SC Museum in Columbia, SC was also instrumental in getting me interested in developing a history of the Clemson Aero Club from its beginning and how it has evolved over time.

The book is arranged in a series of chapters that cover in most cases, time periods of 10 years or less with supplemental information on specific time periods, when more information of the clubs operation was available. A number of photos are included to give the reader a visual history of the information presented. In some cases there are details on specific events that some of the members remember in their lives while a member of the Club.

Otha H. Vaughan Jr.
Huntsville, AL
2007

ACKNOWLEDGMENTS

These individuals are the important contributers to my Clemson Aviation and Space Heritage Book. Without the input from these people who responded to my request for information for this book it would not have be possible. I am very grateful for their contributions in providing the original history for my research in trying to get the information as correct as possible about the Aviation and Space Heritage of Clemson University.

Without the help of Claudia Martin, Clemson University Library Reference Assistant, and Dennis Taylor, University Archivist, who helped me locate and retrieve many of the photos of the Aero Club members, I would not have been able to include them in this history. The many pictures came from the Clemson University photography collections and the TAPS.

The USGS resources were used in locating an aerial photo of the Seneca airport, photos of the first Clemson-Central airport, the Cherry Bottom airstrip, the Oconee Clemson airport in the early days, and photos of the present Oconee Regional airport as it is today.

I must also acknowledge the three people who read and and provided the special reviews on the the various versions that were developed, these being Douglas Mauldin, a dear friend, and former Greenville News Reporter, who I grew up with in Seneca, SC, Dr. Edward Byars, Dr Jerry Reel, and to Dianna Sheran who critically reviewed the next to final version for me and provided much needed editorial information for me on how to improve this document.

1. Joe Rostron
2. Andy Applegate
3. John E. Harper
4. Euel Shelly
5. Dr. Richard Figliola
6. Diana Sheram
7. Dr. Edward Byars
8. John Beans
9. Tim Lewis
10. Jeff Reece
11. Kevin Redmon
12. Lt Col. Katherine Ann DiNuovo
13. Louis Jordan
14. Scott Bowen
15. Warren Jeffords
16. Terry Leape
17. Heyward Douglas
18. Shane Valentine
19. Dr. Randy Collins
20. Paul Weitz
21. Miles Watkins
22. Herbert J Morgan
23. James Jackson
24. Stacy Yikes
25. Walter Mooreman
26. Markus Hess
27. Cole Bristow
28. Gene McCoy
29. Loyd Martin
30. Bill Jones
31. Emory Bishop
32. Jennie Miler
33. Greg McDowell
34. Nick McNair
35. Claudia Martin
36. Dennis Taylor
37. Mike Hugg
38. Paul Werts
39. Stanley Dubose
40. Ramsey Hawkins
41. John J. Weeks.
42. Caroline Burley Miller
43. Peter Donald
44. Kim Cobb Gramling

45. Max Wigington
46. Carlisle Booth
47. Dr. Mike Smith
48. Daneilo Ramirez Lazzilo
49. Marvin Burley
50. Milt Turner
51. Kieth Parris
52. Jeff Treece
53. William York
54. Bonnie McClure
55. Edgar Hutto, Jr
56. Jeff Reece
57. Jack Nisbet
58. Cookie Fitch
59. Doyce Graham
60. Steve Coons
61. Jack Spillers
62. B.J.Todd
63. Gains Liner
64. Don Turk
65. George Todd
66. Naomia Harper
67. Edward Schafer
68. Bob Russell
69. Dan Holden
70. John D. Leaphart
71. Kevin Prufer
72. Anthony Meldeau
73. Tom Squires
74. Natalie Garrison
75. Tony Gondola
76. Jennie L. Sanders
77. Jacque Gray
78. Allan Schafer
79. Douglas Mauldin
80. Dan Schmiedt
81. Mike Frazier
82. Lytle Brice
83. Roger Wilson
84 Joe Blandford
85. Lt Col Brad Mc Coy
86. Lt Col Frank Rogers
87. Col Cristopher Ayress
88. George Venturella
89. Lt Col Jeffrey Lowery
90. Thomas Ressnor
91. Dr. John Hroumi
92. Col Herry McGill
93. George Moore
94. Kathleen L. Hagan
95. Anthony L. Adbel
96. Major Robert Childress
97. Don Frazer
98. Douglas Bradberry
99. Mike Smith
100. Alyssa Mander
101. Garland Brewer
102. John Bailes
103. Wayne Rholetta
104. Sharon Glasgow
105 Robert Banks
106. Cheryl R. O'Brien
107. Whitney J. Stringfield
108. Susan Hoit
108 Col John L Lephart
109 Dana Anderson
110 Tom L. Cox
111 Joe Gault
112 Fred Cone
113 Mickey Dorsey
114. St John Courtenay
115. Dennis Page
116. Melinda Brown
117. Cameron Blazer
118. Emory Bishop
119. William N. Jones
120. Gary Dubose
121. Burgess Howell
122. Col. Sandy Edge
123. Dawson Luke
124. John E. Harper
125. Jack Sherard
126. Stewart Morgan

CHAPTER 1

IN THE BEGINNING
1888 - 1929

Clemson College, located near Clemson, SC was established on Christmas Eve 1888 by a joint resolution the House and Senate of South Carolina who met to ratify an act accepting Thomas G. Clemson's bequest to the State of South Carolina for the establishment of an Agricultural and Mechanical College. However it was not until November 1889 that after the U.S. Circuit Court of Appeals ruled that in May 1889 that Clemson's will was valid and later Governor John P. Richardson in November 1889 signed it into law accepting Clemson's money and property. Funding for the college's buildings and their maintenance was provided by legislation. Additional funding came from Federal Funds(the Morrill Acts of 1862 and 1890, the Hatch Act of 1887). Additional funds were provided by a one -time appropriation of $15,000 by the SC general assembly and an annual appropriation coming from a tax on fertilizers. Convict labor from the state penitentiary were authorized and did the required manual labor for the construction of the campus and buildings. Clemson was one of the first land grant Colleges and the Clemson board of trustees requested the War Department to assign a military officer to be the Commandant for establishing a military style of discipline for the welfare and training of the Cadets.

Lt Thomas O. Donaldson was the first Commandant of Cadets. He graduated from Patrick Military Academy and West Point and then served in the 7th Cavalry, serving in two engagements against the Sioux Indians. His son John O Donaldson, although not a Clemson graduate but a graduate of Furman University of South Carolina and Cornell University of New York later distinguished himself in aviation combat as a member of the British Royal Air Force Number 32, Squadron shooting down 8 enemy aircraft during the period of 3 July, 1918 thru 1 September, 1918. He was shot down and was later captured after attempting with another British pilot to steal a German two seater observation aircraft. During this attempt he was stabbed in the back by his German guard, however he managed to escape his captors. Later he was captured again but again he managed to excape. He and the other pilot returned to England just as the War ended. He received a Captain's commission in the US Air Service for his combat record although not actually flown in combat as an US Air Service Pilot. The Greenville Army Air Field was constructed in 1942 in support of training of combat crews for the B-25 "Mitchell" medium bombers and later the training of B-24 "Liberator" heavy bomber crew. The field was deactivated in 1946 and was reactivated 16 October 1950 due to the Korean War. The field was renamed Donaldson Air Force Base in 1951 as a tribute to Lt John O. Donaldson

Now let us return to the story of the development of Clemson's Aviation and Space Heritage.

The Clemson Aero Club had its first beginning in 1927 with a small group of eight Clemson A & M College students that banded together to build and fly their own aircraft. Because Clemson had no Aeronautical Engineering Department, the students and their Wood Working professor decided they would try to build and fly their own aircraft. Most of the members had had some experience either building model airplanes or they had been associated with aviation in some form or another before coming to Clemson. Dallas Sherman was an airplane model builder and when he was 11 years old he worked as a barnstormer's assistant in Raleigh, NC. His main job was to get fuel for the barnstormer's aircraft from various gasoline stations that they could locate as they flew to the various North Carolina county fairs. They would place a sign on the aircraft to show who had provided gasoline for their operation. To learn something about real airplane construction Dallas and a friend William F. Gaffney also helped rebuild a wrecked aircraft there in Raleigh. He also visited the Ford Airplane plant and attended the American Air Show in Washington, DC. There were initially ten members in the group but it was narrowed down to eight members. These eight members were Dallas Sherman, Charlisle Booth, Nathaniel Green, Hooks Johnson, William Thompson, John L. Marshall, and the Mood twins, Norwood and Preston. They would become in a true spirit the "Flying Tigers".

The catalyst for the club was Professor John L. Marshall. He was Head of the Workshop. Marshall was a graduate of Clemson Class of 1909 and after working on a farm and then also working for Western Electric in Chicago, Ill. he decided to return to Clemson. So in 1917 he returned to Clemson to be a shop work instructor. He really liked his job and was extremely proud to be a teacher. He was recognized by the Clemson students and the college administration as an outstanding educator and in 1919 he was appointed Head of the Wood Shop. He was a great teacher in that his teaching was full of wisdom and concern for his students and their future careers. He was instrumental in the formation of the Tiger Brotherhood Society established to help preserve and uphold the ideals of Clemson. He also promoted the Honor system at Clemson. He was full of energy and was always ready to help the students with their required projects and with other non college projects as in the building of an airplane.

Because students had little money to finance the buying of materials for the project, each member contributed thirty five dollars to the project, and scavenged whatever else they could find to use for the project. The students looked at several designs and finally settled on one similar to the airplane that was designed by Ed Heath called the Heath Parasol. The Heath design was a popular one at the time with the fuselage being welded steel tubing with bracing. So this type of construction was eliminated from consideration and the students decided to construct their aircraft from wood and fabric. Their aircraft was of a parasol design with a wing span of 23 feet and a fuselage 16.7 feet long. The aircraft had a fabric covered wing, fuselage, horizontal stabilizer, elevator, a vertical stabilizer, and rudder. The wing came from a wrecked WWI fighter aircraft, a Thomas-Morse S-4C Scout. Charlisle Boothe repaired a number of the wing ribs as well as constructed new ones as required. The fabric for the plane's skin was donated after Dallas saw it during a Greenville Textile Exhibition that was being held at the College and decided that it would be an excellent material to use. The fuselage was constructed of a series of wood formers with stringers and wire bracing with turnbuckles to provide the necessary strength for the fuselage. The propeller was six feet in diameter, hand carved, and made by laminating five layers of oak. Each tip was covered with a cloth sleeve, held in place with glue, to keep the tips from splitting. The design of the aircraft was such that the pilot's cockpit was located at the aircraft's center of gravity which added to the stability and control of the aircraft.

The engine, given to the students by the Mood's grandfather, was a Lawrance 2 cylinder Model A-3 which developed about 24 to 28 horsepower operating at 1400 to 1450 rpm. It had two horizontal opposed cylinders and a single throw crankshaft. This caused the engine when running to have a lot of vibration because both pistons were moving in the same direction at the same time causing excessive lateral loads. The engine was originally designed for use in the Breese Penguin aircraft, a non flying ground trainer that was developed during WWI. The pilot trainees used a small aircraft trainer to learn how to handle the aircraft on the ground as they taxied it around. The little engine had a bore of 4.00 inches and a stroke of 6.00 inches resulting in a displacement of 150.8 cubic inches. It weighed 128 pounds. The ignition system consisted of a dry cell battery, magneto, and a spark coil that provided the ignition source for the two spark plugs. "It was more like a door bell ringer," remarked Dallas Sherman. The metal fuel tank was fabricated by Hooks Johnston, an experienced welder of the group. It was mounted on top of the wing.

By working on weekends, between classes, and any other time they could devote to construction, the little aircraft was finally finished during the winter of 1928-1929, a high wing, one seat monoplane. They then applied for an aircraft number so it could be flown. After completion of their aircraft the students loaded it on a truck with the help of some College personnel and transported it to the Greenville SC Airport.

In March 1926 Knox Haynsworth, W.H. Allen, and Sam J. Taylor formed the Alhayor Company, the first commercial aviation firm in the State of South Carolina. The company made aerial photographs for use by real estate developers. The Greenville Airport was located on the Cedar Lane Road area of the City of Greenville, SC. About the same time the CAA became interested in an airmail stop between Atlanta, Ga. and Greensboro, North Carolina but the city council did not appropriate funds for the airport so Spartanburg, South Carolina became the official stop over. George Barr, former Army Air Corp pilot and Errett Williams, a local barnstormer, began to lobby the city fathers for an airport at Greenville. By 1928 the legislators created the Greenville Airport Commission and a tract of land was purchased near the Laurens Road on September 9, 1928. The formal dedication for the airport took place on September 9, 1928. The early Greenville Airport was located on the Cedar Lane Road in 1926 but by 1928 the State Legislators created the Greenville Airport Commission and land was purchased near Laurens Road and it was officially dedicated September 9, 1928. This airport is the airport from which the "Little 372" was flown in it's initial flights.

The Clemson students submitted earlier the required paper work during the winter of 1928 for the aircraft to be officially recognized as an airworthy aircraft by the Federal Government. The first attempt to register the aircraft came about as the formation of the Federal Air Commerce Act of January 1927. This Act was created to establish a system of class letters that designated what type of aircraft was being registered. The classes were C, S, or P(Commercial, State, Private). However in 1929 a new plan for aircraft being registered stated that the aircraft being registered would have three numbers and with a suffix of the following: E, H, K, M, N, V, W or Y and Gliders were identified as G. The Clemson Aero Club aircraft was assigned the number N372. Now it could be test flown.

Professor Marshall and Dallas Sherman decided that it would be better if the aircraft would be test flown by an experienced pilot and not by the students. The aircraft was transported to the Greenville, SC airport during the 1929 spring holidays and Williams Flying Service personnel located at the airport agreed to do the test flights. The first pilot to pilot the aircraft was an experienced pilot named George M. Keightly. George had flown in the Mexican Revolution Fight as one of the American mercenary pilots to fly on the side of Francisco "Pancho" Villa. Keightley had some twelve years experience as a pilot. Another Williams pilot who flew the aircraft was Heiek. Keightley stated, after his test flight that the aircraft was of good design and flew well but it needed a more powerful engine to get better performance. The aircraft was well designed in that the pilots seat was located at the center of gravity of the aircraft.

A number of other pilots flew the "Little 372" but in most cases they never got it much higher that 15 to 20 feet above the ground and were not impressed with its performance. The little aircraft however flew better that the builders thought it would.

Because of its small engine the little aircraft required a long take-off roll but once airborne it could cruise along at a speed much faster than the designers had expected. The aircraft was flown a number of times from the airport there in Greenville but because the engine was not very reliable and with its bad vibration characteristics it caused some structural members to be damaged. It was decided that that the aircraft should not be flown any more so the club members decided it would be better to give it to the college and let it hang from the rafters in the woodshop where it was built.

The pioneer Clemson Aviators went after graduation from went to careers in aviation and other fields. Dallas Sherman graduated in 1929 in Architecture and went to work in Pittsburgh, Pennsylvania. In 1932 he entered active service in the U.S. Army Air Corps to begin his training as a pilot, graduating in 1933 as a military pilot. He flew a number of airmail routes as a military pilot when the Army Air Corps was requested to deliver the mail because the civilian contractors were having difficulty accomplishing their contracts. He left military service in 1936 to work for Superior Oil Company of Houston, Texas as a pilot and a member of the Texas Air National Guard as well as other jobs as required. During 1939 thru 1941 he attended Law School at the University of Texas in Austin, Texas and in October 1941 he joined Pan American Airways (PAA) as a check pilot. He later was stationed at PAA Operations in Accra, Gold Coast, Africa. In Africa he flew as a PAA Clipper Captain and while there he was called to active duty, reporting as a Major in the US Army Air Transport Command when the Airlines were mobilized. After four years of active duty in Europe, Africa, and the Middle East he was released from active service as a reserve officer and returned to the USA. Then in 1946 he rejoined PAA in their Washington, DC office as Systems Operations Representative. He was then transferred to the New York Office. While there he became PAA Special Executive Representative, Orient with Headquarters in Tokyo, Japan where he worked closely with the US Government helping with the economic rehabilitation of Japan after the War.

One interesting event during WWII in which he participated was a special military flight with another Clemson Mechanical Engineering graduate, Major Theodore J. Boselli. Dallas was the Senior Pilot and Boselli was the Navigator for a "Pathfinder Flight". This flight by the "Sacred Cow" duplicated the proposed flight path that President Roosevelt and his supporting staff were to fly to the upcoming Yalta Conference. After the "Pathfinder Flight" Theodore was assigned later to be the Navigator for the President's flight. During his military career Theodore was the navigator for many of the flight of dignitaries from Washington National Airport. Some of these dignitaries were Ambassador Averill Harriman, Ambassador William Bullitt, The Truman Committee (Senators Henry Cabot Lodge, A.B. "Happy" Chandler, Richard Russell, Ralph O. Brewster, and James M. Mead), General George C. Marshall, General Dwight D. Eisenhower, President Harry S. Truman, Madame Chiang Kai-shek, Winston Churchill., Dr Edward Benes of Czechslovakia, and General Skorshi of Poland. He later flew combat tours over Korea and before retiring as Col from the USAF, he was logistics officer for the 15th Air Force.

Hooks Johnston went to work for an aircraft engine manufacturer in Connecticut. He did a little barnstorming for a while and later in his career became the Supervisor of Engineering Assembly Quality Control for the Pratt Whitney Aircraft.

Preston Mood continued to fly and in 1938 joined American Airlines(AA) and later became a Captain for AA. During his career with the airline his seniority number was 268. He and Ernest K. Grann, the famous aviation writer trained together when he first applied to AA for a pilot training position and they later became great friends. Mood flew over 100 flights over the Atlantic and Pacific Oceans. His brother Norwood, also a pilot, became an FAA Flight Inspector for the Southeastern Division.

Nathaniel Green, after a distinguished military flying career in the Orient during WWII and other military flying assignments was assigned to Shaw AFB, Sumter, SC, later retiring there as a Major.

William A. Thompson after graduating from Clemson went to work for the Chamber of Commerce at Sumter SC. While he was there he was instrumental in getting the War Department to establish the present day Shaw AFB, in Sumter, SC. He entered the Army Air Corp and later became the first executive officer at Shaw AFB, at Sumter, SC. After leaving the Air Corp he became one of the top salesmen for the Williams Furniture Company of Texas.

Charlisle Booth graduated from Clemson Class '32 with a BS Industrial Engineering and later received a Masters Degree in Aeronautical Engineering from North Carolina State University. Later in his career with Bureau of Aeronautics Avionics Division, Washington, DC he became an assistant to the Secretary of the Navy.

So it seems that the Aero Club had quite an influence on the lives and future careers of its first members.

CHAPTER 2

1929 - 1939

H. D. Martin was elected president of the Aero club in the fall of 1929 and the club members agreed to repair a primary glider that was built in the woodshop by students and had been damaged on one of its flight earlier that year. It took about three months of effort to build the glider. This single seat glider was also registered with the government later on May 28, 1930 as Clemson Aero Club G-1 N10276 and was still on their records as of July 30, 1934 when it was deregistered.

On April 22, 1932 the club members purchased a damaged Eagle Rock biplane. This aircraft, a 1928 Model A-2, Serial Number 58, with an OX-5 engine, had been registered with the government as 7194, Serial Number 58. The students were planning to repair it to be used as a training aircraft. It was repaired and may have been flown by some of the Club members. The club members that purchased the aircraft at this time were H.D. Martin, J. B Chambers, H.C. Woodson, H.L. Wyatt, and Col Howard Behr. The aircraft was sold on June 24, 1934 to Joseph N Musleh of Jacksonville, Florida.

They also acquired a Le Rhone 120 horsepower aircraft engine that they used as an experimental power plant. The interesting thing about this engine was that the engine cylinders rotated with the propeller and that the engine was controlled using only an on-off ignition switch.

Around 1934 there were still some new students that entered Clemson that were interested in aviation and one of theses students was James D. Nisbet. He became interested in flying when he earlier witnessed Capt Elliott White Springs, a World War I Ace of Fort Mill, SC, fly under the Buster Boyd Bridge earlier. He and six or seven Clemson students decided to form a new Aero Club. After discussions with Professor John L. Marshall, the mentor and faculty advisor to the original club, he agreed to work with the new club members. He told James about a former military pilot named George M. McGee who had a flying operation at the nearby Anderson airport. George had a Taylor E-2 Cub that he used for training and James and the other students contacted him to provide an aircraft for their club's flying instruction. Jim became the president of the club and one of the club's members was a student named Lindberg and another student was named Barney as Jim Nisbet recalled in his book "Wheels Up". Jim soloed at the Anderson Airport after two and one half hours of instruction in the Cub . After several members of the Club had soloed George bought an Aeronca C-3 aircraft that could carry two people side by side and had a two cylinder Aeronca E-113 engine that developed 36 hp. George moved his aircraft to an abandoned golf course near Clemson that had closed during the depression and began his operation. The Aero Club used this location for a while until Nisbet and the other students were graduated.

During the club's operation they acquired a World War I V-12 Liberty aircraft engine and test stand and during his junior year Nisbet decided to demonstrate the Liberty engine by running it with his professor's permission during Engineering Week. With the help of the other aero club members and some other Clemson maintenance personnel he was able to move it to the valley behind the Engineering Building where it was cranked up. It was said that during the second day of Engineering Week the President of Clemson told them to stop running it as it made too much noise and disturbed classes. Professor Ben Furnow and Professor James Sams were persuaded by the aero club members to teach a class in aeronautical engineering as part of their mechanical engineering studies. Professors Sams and Furnow's teaching inspired Nisbet and he became really interested in things mechanical as well as aeronautical. Professor Sams, a U.S. Army Air Corps reserve officer, lectured about the internal combustion engine's problem in maintaining their sea level power in the rarefied air of the stratosphere where he believed airplanes would be flying in the future. They would need some method of compressing the air and delivering the air to the engine. The supercharger was the device that could do the job. The supercharger was invented by Dr Morse around 1918 to get the Liberty engine to develop sea level performance at altitude, and the tests that were conducted at Pikes Peak in Colorado were successful. However since the supercharger used a compressor that was being driven by an exhaust gas turbine there were problems. The hot gasses flowing thru the turbine blades, made of stainless steel, caused the blades to creep and lose their shape. Sometimes they would damage the enclosure and cause the blades to fail, thus rendering the supercharger useless. The lifetime of the supercharger was limited due to not having "High Temperature" metals. Solving this type problem would later be a challenge to Nisbet in his future career.

Nisbet graduated in 1937 with a degree in Mechanical Engineering and went to work with the General Electric Company of Schenectady, NY. in their Manufacturing and later in their Research Division. At GE he had a very exciting and challenging career working in a number of positions in the development and manufacturing of transformers, electrical motors and finally in the in the development of high temperature super metals that could be used in superchargers for high flying military aircraft during WWII. The Boeing B-17 Bomber aircraft, Consolidated B-24, and a number of fighter aircraft such as the Lockheed P-38 and Republic P-47 were equipped with superchargers called turbo superchargers. Although they were standard equipment in the aircraft, there were still operational problems with them. Nesbit realized that a new field of research that could provide better metals would be needed in the future development of the supercharger, the newer jet engines, and more efficient steam turbines that were evolving so he decided that he would become a metallurgy trainee there at GE. During his stay there at GE while working in high temperature metals research he got a chance to meet Sir Frank Whittle, the inventor of the British jet engine which would need the better metals to operate more efficiently.

The problem with the production of early high temperature metals was that the use of some materials to create the alloy meant that it could not be melted in the standard furnace. They were so reactive to oxidation to the atmosphere that when produced in the standard melting furnace that the alloy produced had weak grain boundaries and were weak at high temperatures.

After WW II was over and American industries began to produce products for the civilian markets. GE was interested in improving their gas turbine and jet engine products so they decide to create a laboratory for the research, development, and the production of special high temperature alloys and Nisbet was invited to continue his work with GE in this new laboratory as a research metallurgist. While there he was in charge of 15 scientists and he was involved in the development of the vacuum melting furnace that is standard for making all high temperature alloys. He left after 16 very fruitful years at GE and accepted a job as Director of Research and Development for Cylops, a company that specialized in the production of high temperature alloys that were produced using a vacuum furnace. After three years with Cyclops Jim decided in 1957 to form his own company Allvac Metals and built his plant in Monroe, NC with the help of local businessmen. The plant was located in an area which had been Camp Sutton during WWII Around 1962 Jim purchased a large tract of land between Monroe and Charlotte, NC of about 500 acres which he named "Aero Plantation". Here he built a six-bedroom house with a plane port, his own 2500 foot runway, and his own cattle farm. Since Jim and his wife were pilots this was an ideal place to enjoy life while working. Around 1965 the Monroe Airport was closed and so he subdivided some of his land so that other pilots who needed a place to keep their airplanes could build their homes and have a place to continue flying. He later merged with Vasco Metals and finally Allvac and Vasco becoming a subsidiary of Teledyne. Monroe did not have a commercial airport and since there would now be more commercial travel required for the company managers to visit the new Teledyne Titanium plant which was under construction Jim decided to push for a commercial airport there at Monroe. He decided that the airport should be built near the Titanium plant so he then purchased about 60 acres of land for the Titanium plant. He also purchased an additional 400 acres for a proposed airport. He then created a venture-capital company called "Ventures" with a number of partners and Ventures hired the services of an airport architect to design an airport with a 4000 ft runway. This company later presented their design for a proposed airport and with the assurance that the FAA would pay for its development, the city fathers agreed that they needed the airport. Jim now had his commercial airport that he had dreamed of. Jim later served as Group VP at Teledyne Corp until 1970 when he resigned. Since 1970 he has been a venture capitalist, money market manager, a stock market researcher, and the author of 6 books.

Jim's association with aviation in the Clemson Aero Club which he founded in 1934 seemed to have had quite an influence on his life and career.

To determine what happened to the Taylor E-2 that the Aero Club members had taken their initial flight training there at the Anderson Airport, Anderson, SC, I researched a number of sources and found that the airplane was later sold to Carmel City Flying Service. The owner later was approached by a young man, Tom Davis, who wanted to learn to fly and he soloed the aircraft. Later he was hired by the flying service, which sold and leased airplanes, trained pilots and offered maintenance service for the aircraft. He left Winson Salem, NC to attend College in Arizona and study pre-med. Here he was hired by the local flying service to be a flight instructor while going to college. He left College in his senior year in 1939 to returned to NC and began working for his old flight instructor L.S. McGinnis the owner of Camel Flying Service. He later in 1940 helped refinance a loan that the company was struggling to pay and became Vice President and Treasurer for the flying service. He then changed the name of the flying service to Piedmont Aviation, Inc., a holding company for Piedmont Airlines. By 1941 Piedmont Aviation was the first fully certified aircraft and engine overhaul facility between Washington, DC and Atlanta, Ga. that was approved by the Civil Aeronautics Board (CAB). During WWII Piedmont Aviation, Inc trained military pilots and after the war and with the end of government contracts Davis applied to the CAB for a Certificate of Public Convenience and Necessity to operate a local-service passenger, mail and cargo route to the Southeast. Piedmont Airlines was granted on April 4, 1947 the route 87, which would link various points in North Carolina to the Ohio Valley. However the route was postponed due to protests from State Airlines and Eastern Airlines and resulted in the CAB awarding Piedmont certificate number 208 on December 12, 1947. More protest again developed from State Airlines and Piedmont Aviation, Inc and Davis after appealing to the US Supreme Court, ruled in favor of Piedmont Aviation, Inc.

Piedmont Aviation was formed on January1, 1948 separating the airline from Piedmont General Aviation into autonomous branches under the Piedmont Aviation umbrella.

On February 20, 1948, Piedmont Airlines flight 41 departed Wilmington, NC for Cincinnati, Ohio with stops in Southern Pines, Charlotte, Ashville, Tri-Cities, and Lexington, KY. On board was Tom Davis and several company officers, and one paying passenger. The little Taylor E-2 is now registered to Sun N Fly-In, Inc, a museum, now located in Lakeland, Fl. It was given to them by the family of Tom Davis.

CHAPTER 3

CLEMSON AND WWII
1939 - 1945

It was now 1939 and with the War in Europe on the horizon there were a number of people like Col Charles Lindbergh, first to fly solo across the Atlantic to Paris in 1927 and a national hero who believed that America might be drawn into a War if it developed in Europe. In 1936 he was asked by the American Military Attache in Berlin, Major Truman Smith, to come to Germany to learn more about the capabilities of the German Luftwaffer that had been developing over a number of years since the end of World War I. Major Smith believed that Lindbergh's background in the civil and military aviation would be a valuable asset to America in accessing the Luftwaffe's capabilities and would allow him to see things that normally would not be see by other government and military persons. Lindberg after seeing the large number of the Luftwaffe's aircraft that had already been produced and the number of Germany's aircraft manufacturing plants and their efficiency and capabilities he realized that America was not prepared to wage an aerial war if called upon to help her allies. He later briefed the Army Air Force Intelligence people, and testified before the House and Senate and other government officials in opposition to American intervention. He resigned his commission as an Army Air Corps Reserve Lt Col at this time so he could speak as a private citizen about his opposition of America becoming entangled in the European war that was on the horizon. He later gave a number of radio and anti-war public speeches stating that America should remain neutral, build up our military strength, and that we should stay out of the War in Europe which was now developing. At the same time there were many others in the country that believed we needed more trained pilots if a national emergency arose or if we were ever drawn into the conflict that was now developing in Europe. Nell Behr, the first licensed woman pilot in South Carolina, and her husband Lt. Howard Behr operated a flight training school there in the Anderson Airport in Anderson, SC. In 1936 she became a member of the Ninety Nines-a women's pilot organization and later was the Vice Governor of the Southeast for that organization. She began speaking in the National Ninety-Nines meeting around 1939 and she proposed that the government should set up contract flying schools to train pilots if and when America was drawn into the War now developing in Europe.

There were discussions in early 1940, after a visit by Mr. Dexter C. Martin, Director of the SC Department of Aeronautics Commission who met with Clemson's Agricultural College Business Manager J.C. Littlejohn about a proposed landing field to be constructed near Clemson. Clemson's Civil Engineering Department under the direction of Dr. C.E. Glenn and a number of his students were tasked to prepare a topographic map of the proposed site for a landing field. They also met with the official in charge of the Civilian Conservation Corps (CCC) and they were informed that the CCC would be glad to co-operate with Clemson in the establishment of a landing field. By April 16, 1940 the topographic map showing the profile and cross section grade lines were completed. These data were delivered to Mr. Martin for his consideration. He was shown a design for the construction of the proposed landing field with two runways. Engineers from the SC Aeronautical Commission after their review stated that the proposed site was not suitable but that a site in the Cherry Bottom area about three miles south from the Campus could provide a landing field with 3 runways of at least 3500 feet in length. On April 10, 1941 the South Carolina Council for National Defense Director Heyward Mahon corresponded with Mr. Martin in reference to the new program for the construction of military airports, however the military airport was not constructed at the Cherry Bottom site.

Clemson students since early 1939 had been receiving aviation flight instruction and ground school training under the Civilian Pilot Training Program (CPTP) that was created by Congress and signed into law by President Roosevelt. Mr. Martin in a letter to Mr. Mahon stated that he would like for an airport to be built at Clemson but since the Clemson students were already receiving their flight training at the Anderson Airport it would be better to improve the Anderson Airport rather that build an airport at Clemson.

Clemson was one of the 250 schools selected to provide civilian pilot training under this program. Clemson College announced on October 5, 1939 that classes for the CPTP would be held under the Direction of Dr. Sam Earle of Clemson's Engineering School. Of the students that participated in the program that started in the fall of 1939, seventy five students applied for the classes and forty were selected after they had passed their flight physicals. The students were called the Clemson Flying Cadets(1939 -1942). They were to be trained in their ground school by members of Dr. Earl's faculty and their flight training was to be conducted at the Anderson Airport by Lt Howard Behr, the manager of the Anderson Airport, initially under a government contract. Later Southern Airways, Inc. and its instructors under the management of Lt. Ike Jones, who held a commission in the Air Reserve, became the government contractor for the CPTP at the Anderson Airport. Dr. F. H. Calhoun of the Chemistry Department taught Meteorology, Dr. J. F. Sams taught the student pilots the mechanics of flight, and Ben Jordan, assistant superintendent of buildings and grounds and a mechanical engineering graduate of 1938 taught the students the Civil Air Regulations. The first 6 students to complete the course were all engineering students, two were Electrical Engineering and the other four were Mechanical Engineering students. After successful completion of then ground school and their flight training they were awarded a Private Pilots license and if they desired they were allowed to take an advanced flying course in aerobatics. The first six students that completed their flying and ground school and received their private pilots certificates were: J. L. Martin, E.P. Manville, C. B. Pool, Bob Taylor, R. R. Willey, and R. R. Agnew. Shortly after Agnew entered military service in 1940 and after further training he began flying as a military pilot. He and a passenger were killed in the crash of a military aircraft in November 1941 while flying a cross country mission in California. The accident according to investigators was probably weather related.

One instructor for Southern Airways was Edward T. Shafer. He started his aviation career in the late 30's. He received his instruction from Bevo Howard and Martin Jensen at the John's Island Airport, Charleston, SC. While a student at Clemson he continued flying and received his private pilot license in 1938, and later his commercial and instructor license in 1939. He entered in the Civilian Pilot Training Program and was employed by Southern Airways, Inc as an instructor, instructing at the Anderson Airport in their CPTP operation. There were four instructors and about 40 students. The students were trained in J-3 Cubs powered by 50 hp Franklin engines. His initial students were Cadet C. B. Pool, Cadet Reid (he later became a Pan Am Captain), Cadet Embree, Cadet Darby, Cadet McLeod, Cadet C. L. Harrington, Cadet London, and Cadet Farmer. They were all awarded their Private license. He was later assigned to the Greenville operation in June 1940 to train CPTP students in the Furman University program. Later in 1940 he joined United Airlines flying the legendary Boeing 247 and then Douglas DC-3 aircraft. He later joined Pan Am in 1942 and served as a Captain on the Central and South American Routes. He flew the inaugural Boeing 727 flight from Berlin, Germany to Frankfort, Germany in 1965, breaking the ground speed record for a flight between the two cities. During his 43 year career he logged in excess of 30,000 hours. He served as a committee chair for the Airline Pilots Association (APLA) for more that 10 years. He is still involved in flight instruction at the age of 83 and was inducted into the SC Aviation Hall of Fame in 2003.

Miles Hunter Watkins participated in th CPTP while he attended Clemson during 1939, 1940, and 1941. He first soloed a J-3 Cub. After completing his ground school and flight training he was awarded his Private Pilot Certificate on September 15, 1940. After completing this phase of training he elected to take the Secondary Course. In this phase of training he learned aerobatics and other specialized training while flying a Waco UPF-7. He completed this phase of his advanced training on June 15, 1941. A typical cross country flight as required for the Private license consisted of a take-off from the Anderson Airport, then to the Camden Airport with a full stop landing, and then proceeding to the Columbia Airport with a full stop landing. Then fly back to the Anderson Airport. Miles after graduation in 1941 and receiving his reserve commission, was called to active duty June 10, 1942, and remained for the duration of WWII. More details about Miles can be found in Chapter 13.

Dr. Sams, a reserve Army Air Corps office, taught till 1942 when he was called to active duty with the U.S. Army Air Corps and served for the duration of the war. The CPTP continued till late 1943 and was terminated and restructured eventually under the direction of the U.S. Army Air Corps. A number of the Clemson students distinguished themselves as pilots, navigators, bombardiers, and/or other types of aircraft crew members and or ground crew members during WWII. Some of these students' biographies are presented in Chapter 13.

During February 1942 the local unit of the South Carolina Defense Force at Clemson became the Headquarters for the South Carolina Defense Force and about 400 individuals registered for civil defense work. Clemson College began the training of 30 people of Company G, 2nd Regiment of the South Carolina Home Guard located in Seneca, SC. in the use of automatic weapons (Browing Automatic Rifles and the machine gun).

The School of Vocational Education and the Engineering Department of Clemson offered government sponsored courses in auto mechanics, forge and foundry, and welding with night classes being offered to any individual that wanted to support the War effort. The School of Engineering offered Engineering Drafting, Radio Fundamentals, Safety Engineering, Surveying and Mapping there at Clemson as well as in Anderson, Greenwood, Greer, Spartanburg, Walhalla, and Seneca. My father enrolled in the Engineering Drafting Class there at Seneca, completing it in late 1942.

With many students being called to service in early 1942 and not being able to complete their requirements for a degree, the trustees of Clemson College authorized the granting of a degree if the student had completed seven semesters with an exemplary record and were drafted.

In the fall of 1942 Army paratroops were stationed in the Clemson area and used the Clemson College lands for practice jumping operations and also they used the ROTC Rifle range.

In mid November 1942 Colonel Herman Beukema, a West Point Graduate and former professor of history and government at West Point, was selected to become the Director of the Army Special Training Department (ASTD). The Army Specialized Training Program (ASTP) which had been under consideration for some time now became part of the larger program. The ASTP was offered in two phases, one of which was much like the first one and a half years of college and the second phase was much like the last half of the sophomore year at college. This training could help provide the Army with specialists that were needed but could not be trained at Army facilities as they were not available.

In early 1943 Clemson was one of 281 colleges to provide housing and educational facilities to the U.S. Army Air Corps for their pre-flight training programs for their cadets.

On March 4, 1943 the first personnel of the Army Air Corps Aviation Cadet trainees arrived at Clemson for their initial pre flight training. The training consisted of academic, cultural, physical training, and disciplinary training that would prepare them for their Officer's commission if they successfully passed their flight training that would be offered to them in the future. They also received ten hours of primary flight training instruction at the Anderson, SC airport. After five months of training they were sent to a classification section at Nashville, Tennessee and then later to a pre-flight school for 3 more months of training. About 100 trainees arrived and departed from Clemson each month so there were approximately 600 trainees at Clemson at any one time. One medical officer, three medical non-commissioned officers, five Army Air Force officers , and eight enlisted men were assigned to support the operation.

By the 1943 spring semester there were about 1600 students and 304 graduating seniors. The entire rising senior class was called to active duty as privates and by December 1943 105 of these graduating seniors had completed Officers Candidate school and were commissioned in one month.

In early May 1943 Clemson received the first of the Army Specialized Training Program (ASTP) of about 500 that came to Clemson for specialized College training. This program was to be a nine month accelerated learning program for specialized military personnel that had some earlier college training and that were already serving their country. There were now 16 full time officers and 16 enlisted men assigned to help the Clemson College Administration conduct this program.

There was also at Clemson a Specialized Training and Reassignment(STAR) program. This program was used to determine the aptitude and qualifications for young men where they took psychological and classification tests, they answered questions about themselves, their education level, and had discussions with various professors that might be teaching the courses that they could specialize in while there at the college. After completion of the testing they were assigned to other colleges which taught those subjects that they were interested in or they were sent back to their respective units.

The Clemson Civilian Conservation Project forest area (about 135 acres) was selected and designated as a land and water bombing range for use by the North American B-25 light bomber aircraft that were stationed at the Greenville Army Air Base, Greenville, SC, later renamed as the Donaldson AFB in 1951. There were two bombing ranges in the project area, these being a large land target area and a water floating target located on Isaqueena Lake. The B-25 aircraft used a large flag like canvas target with a big red bull's eye. The target was mounted on a floating platform in Lake Isaquenna and represented a ship. The B-25 aircraft used Tilman Hall building on the campus of Clemson College as their initial point (IP) for their bombing run on the land target range. They used it also as their IP before dropping down to a lower altitude for their initial attack altitude from which to begin their skip bombing run. Skip bombing was a way to attack ships and was used effectively in the South Pacific theater of operations to sink Japanese ships. In skip bombing, the bomb after release from the aircraft hits the water and then it skips off the water, bouncing, and exploding in the ships hull above the water line.

In the summer of 1943 there were 600 Army Air Force trainees, 400 regular students, 500 basic trainees, and about 800 to 900 STAR students. During this time the official cadet uniform for the students was the Army khaki uniform with distinctive ROTC insignia to help distinguish the cadets from the military personnel. During this time the cadets were housed in separate barracks from the Air Corps and ASTP, and STAR students. During the fall of 1944 enrollment there were 957 students including 38 veterans that were attending college. Thirty five advanced ASTP students still in the program received their GI Diploma in Electrical Engineering. In August a final group of Army Air Corps Enlisted Reserves (ACER) arrived with 300 students for the summer semester and later in November 330 ACER students arrived for their training. The ACER students were considered as part of the ASTP but they received no pay while they were in inactive training.

The ASTP program as mentioned earlier was created by the Army to provide students that had completed two years of their college education a chance to finish the college education and then they would be offered officers commissions after completion of the program. However in early 1944 the program was terminated due to the Army's increased manpower needs to support the D-Day Invasion and other Army manpower needs in the Pacific. The students were then sent to regular Army Commands as needed.

When the ASTP enrollment in early 1944 was reduced because of manpower shortage, students from the Army Specialized Training Reserve Program (ASTRP) that were under 18 years of age and not subject to the draft were sent to Clemson.

The best source of information about the Army Specialized Training Program (ASTP) is detailed in the book "Scholars in Foxholes" by Louis E. Keefer. He points out that the Army during WWII conducted the largest college education program in our nations history. Those men that volunteered, somewhere around 200,000 soldiers, were assigned to 227 different Colleges to participate in a highly specialized educational training program. These students received training in various engineering, medicine, dentistry personnel psychology, and also there were 34 different languages that were taught. The massive arrival of these military personnel overnight caused many campuses to resemble Army training installations as Keefer remarked in his book.

Requirements to be in the ASTPwere to have a high school education with an ACGT score of 110 to qualify. Later this score requirement was raised to 120. The average age for most trainees was 18 to 21 years but older personnel were accepted if they had some college credits.

The standard work week was strictly military--59 hours of "supervised activity" which included a minimum of 24 hours of class work and lab work. 24 hours of required study, 5 hours of military training, and 6 hours of physical fitness training. There was still some time for socializing with the local people at the colleges. According to Col Herman Beukema, who came to the Pentagon to run the program and was a highly respected, long time professor of history and government at West Point, said that the ASTP was more rigorous than those at West Point or at the Naval Academy. By Christmas 1943 there were some 140,000 men in the program and this was a high point, but in less than 2 months later , the WAR Department on February 1944 unexpectedly announced that 110,000 ASTP members would be returned to line duty by April 1, 1944. Some 35,000 in special advanced schools were allowed to continue their education. Many of the ASTP members however were sent directly to infantry, airborne, and armored divisions some still stateside in training before the divisions were deployed overseas. A large number of Army Air Corps cadets were dropped from the program and they were assigned to the infantry due to manpower requirements at that time.

To paraphase Keefer-"The life of an ASTP member was not a bed of roses as a lot of old army regulars considered the green ASTPer to be just a smart ass bunch of college kids that wanted to steal their stripes and needed to be taught what the 'real' Army was all about". The ASTPers themselves didn't help themselves as they stuck together and this caused friction between the old guys and the greenies.

Once in combat however these prejudices were soon forgotten and many ASTP members turned out to be good soldiers where ever they were assigned. Many of these gifted and educated men were lost on the beaches on D Day and in the other battles in Europe and in the Pacific. So America lost a lot of technical expertise in WWII that would be needed at the end of WWII. However the ASTP was a catalyst that led many of the returning veterans to finish their education and many distinguished themselves after graduation by becoming Physicians, Engineers, Lawyers, Senators and Congressmen, Teachers, Professors, and other professionals.

A number of former Clemson ASTP members came back to Clemson to continue their education. One of these was John D Hromi.

DR. JOHN D. HROMI

Educator, Practioner and Consultant in the Field
of Quality Control

John D Hromi, after being tested for placement in the ASTP by the US Army at Auburn University in 1943, was sent to Clemson to be an ASTP student at Clemson A & M College and during the eighteen months there he completed some of his studies toward an Electrical Engineering degree. He was then assigned to a radio repair school at Camp Crowder, MO while other ASTPers from Clemson were sent to other staging areas and then to the European Theather of Operations. He entered the Army Signal Corps Officer Candidate School and received his commission as a 2nd Lt. In 1946 he was released for service but remained in the Army Reserve. During 1948 he graduated from Carnegie Tech with a BS in Mathematics. He was invited to return to Clemson and become a teacher in the College of Engineering, teaching statics, dynamics and strength of materials with lab. Professor Rhodes, Head of the Electrical Engineering Department evaluated his credits earned during his ASTP stay at Clemson and determined that he was within a few credits of graduation for his BS EE degree so while a teacher he was also a student. He complete his studies and was awarded his BS Electrical Engineering in 1950 while teaching from 1948 to 1951. Later he was awarded a M. Litt (Mathematics) during 1951 from the University of Pittsburgh. He continued his studies at the University of

Detroit and in 1971 received a Doctorate in Engineering. He held positions at Lawrence Institute of Technology as Chairman and Professor of Mechanical Engineering. He was employed at Ford Motor Company as Principal Engineer in the Environmental and Safety Office and as Manager Engineering Methods Department of the Engineering Staff. He is known internationally as an educator, practitioner, and a consultant in the field of Quality control. He is Professor Emeritus and Founder of Rochester Institute of Technology's John D Hromi Center for Quality and Applied Statistics.

Another student who attended Clemson during WW II was Lyle Brice.

LYTLE BRICE
Mechanical Engineer and Space Pioneer

Lytle entered Clemson in the fall of '41. After Pearl Harbor, December 7, 1941 he signed up for the Enlisted Reserve Corps along with many of my classmates. They were told that when the military was ready for us, we would be called into service. They were allowed to finish their sophomore year and were called into service as a group of about 200 reporting to Fort Jackson at Columbia, SC. After he was called into service, he was sent to aircraft maintenance school at Amarillo, Texas and then to Chanute Field, Illinois and then was assigned to the 58th Bomb Squadron of the 20th Air Force in India about 100 miles from Calcutta. He worked on the flight line doing aircraft maintenance for the B-29 aircraft that were there. He later was stationed at Tinian in the Pacific when the war ended.

He returned to Clemson to finish his education and graduated in '48 with a degree in Mechanical Engineering. He, like a number of veterans who came to Clemson after the War, participated in the Air ROTC program and received his USAF commission as a 2nd Lt. After working at the Tennessee Valley Authority, Dayton Rubber Company and Champion Paper Company for a number of years, he joined the Space Program in Huntsville, Alabama with the Chrysler Space Division. His primary work involved the design of cryogenic propellant loading systems for the Saturn IB Launch Vehicle. He also participated in the design of launch Pads 34 and 37 at the Cape from which the Saturn V's were launched. After the Apollo program was completed he participated in modifications to Pad 39B. From Pads 39A and 39B the launches of Skylab (America's First Space Station). on the Saturn V vehicle and Saturn 1B vehicle(which launched the first crew for the Skylab) and later Apollo/Soyuz Test Project (American-Russian) International mission were launched. After the joint US/Russian program, he joined Reynolds, Smith and Hills in '74. Here he worked in the design of the Space Shuttle launch facilities for NASA. Since '74 he has continued to work in the design of space vehicle processing buildings and launch pads for many aerospace clients in addition to NASA on various space programs.

Another student that attended Clemson during the WW II years was William N. Nickles.

LT WILLIAM NEVILLE NICKLES
WW II B-24 Pilot US Army Air Corps

Lt. Nickles graduated from Murray in the spring of 1941 and attended Porter Military Academy for one year, graduating in the spring of 1942. He then worked in the drafting Department of the Charleston Dry Dock and Ship Building Co. during the summer of 1942. He joined the reserves of the US Army Air Corps in August of 1942. He entered Clemson A & M College in the fall of 1942 and began his study in the field of of Architecture. He was called to active duty in February 1943 and took his preflight training at

DuQueane University in Pittsburg, Pa. He next attended his primary training at Ocalla, Fl. and later began his basic flight training at Bainbridge, GA completing it. Then he went to Advanced Training at Turner Field, Albany, Ga where he received his silver wings as a pilot and was commissioned a 2nd Lt. on June 24, 1944. He then entered his four engine training at Maxwell Field, Montogomery, Al and later to crew training at Tonopah, NV finishing his training in Februaru 1945. He was then transfered to the 5th Army Air Force in the Philipines flying the B-24 Libertator Bomber as First Pilot.

During a series of missions he was cited for meritorious achievement while he and his crew participated in aerial flights from April 23,1945 to June 4,1945 and was awarded the Air Medal. These operations were against enemy installations, shipping, and supply bases.

During a mission to Formosa on June 22, 1945 his aircraft was reported missing in action when his aircraft did not join the formation at its rendezvous point or over the target.

He, like many of the members of the Clemson classes during WWII, gave their ultimate sacrifice for the defense of America during War Time.

When the military training programs officially ended on March 3, 1946, over 10,000 servicemen had trained at Clemson through the various divisions of the College Training Programs (ASTP, ASTRP, ASTP-ROTC, STAR Unit, Air Corp, ERC, and ROTC). In all there were over 6400 Clemson Graduates and former students that served in the military during WWII. There were a total of about 373 former Clemson students and Alumni that were killed during WW II.

Chapter 4

Post WWII Period
1945 - 1950

It was now 1946 and WWII was over and the veterans were now ready to return to the colleges to continue their education that had been suspended by the war. They served their country honorably and now they were eligible for a college education paid for by the government under the program called the GI Bill of Rights. Since Clemson was a land Grant College the cadets were required to wear military cadet uniforms and to participate in military training for their Freshman and Sophomore years but the veterans in most cases were not required to participate in many of the military exercises. The unmarried veterans were assigned to their own barracks and had their meals in their barracks. The veterans that elected to participate in ROTC lived in the barracks with the Cadets and participated in all the military exercises. The day students that were in the ROTC program were required to wear the cadet uniform while on the campus. Some of the married veterans were provided housing on the campus in a military housing units or prefabs as they were called. However some of the veterans wanted to receive their commissions as military officers so they elected to participate in the ROTC program and were active in the military training there at Clemson. The ROTC program at Clemson was conducted by personnel of the US Army and Col A. J. Thackston Jr. was the Commandant. There were 17 Army officers and enlisted personnel and three Army Air Corps personnel, Captain G. C. Salvo, Captain G.W. Hueners, and T/Sgt G. M. Dejanorich that provided military instruction for the future Army ROTC students as part of the Army ROTC program. However on August

22, 1946 General Dwight D. Eisenhower, Army Chief of Staff, signed General Order 124 establishing seventy-seven Air ROTC programs units under the US Army Air Training Command. A few days later President Harry S. Truman formally appointed General Carl A Spaatz to be the first Chief of Staff for the Air Force. Secretary of the Air Force , the Honorable Stuart Symington, who had been sworn in on 18 September 1947, signed a Defense Order on 26 September 1947 that transferred all Army Air Corps Units and personnel including the Air ROTC to the United States Air Force, now a separate military service. These Air ROTC units were placed under the new command, the Continental Air Command (ConAC), which was composed of the Air Defense Command (ADC) and the Tactical Air Command (TAC). This command was created to strengthen the air defense and close air support missions, and placed all tactical fighter resources , including all active, reserve, and guard personnel under a single commander. Between 1948 and 1952 Headquarters ConAC provided military teachers, course curriculum, summer encampment, manuals, and training aids to AFROTC units. ConAC became responsible for what is known today as AFROTC.

General George C. Kenney, General Douglas MacArthur's top air commander during WWII for the southwest Pacific Air operations, became the Commander of the Air University on October 28, 1948 that had been established earlier in 1946. He believed that the AFROTC units should be under the command of the Air University and not Continental Air Command (Con AC) since the Air University offered a more professional and specialized education for officers and non-commissioned officers in preparing them for command responsibilities. He wanted a General Officer in charge of a headquarters for AFROTC who would report to the Air University and be responsible for the curriculum, comptroller, training materials, and operations. He also proposed that there be ten intermediate headquarters commanded by Colonels to manage and control the detachments directly. ConAC did not agree with Kenney's recommendations and the Air Staff then did a number of studies and on August 1, 1952, Air Force Chief of Staff, General Nathan F Twining placed the AFROTC under the command of the Air University.

The Air University now was responsible for the AFROTC and Brigadier General Matthew K. Deichelmann, Deputy Commander for Education was authorized a headquarters and detachments to be located at Maxwell AFB with an overall strength of 1685 officers, 1555 airman, and 29 civilians to administer the program.

During 1947 the Arnold Air Society was formed at the University of Cincinnati by their AFROTC Detachment to recognize their outstanding cadets. The name they chose for their society was General Henry H. Arnold, the first and only five star General of both the US Army and US Air Force. Permission was granted in December 1947 and in 1948 the USAF officially recognized the society and it became a project of the Air Defense Command. They sent out copies of the society's constitution and installed some 20 new squadrons at most colleges and universities during the next few years. At Clemson 30 AFROTC members of the Class of 50 established their Arnold Air Society consisting of 21 Distinguished Military Students and other outstanding cadets. The Clemson Society's Charter was officially signed by some of the members of the 51 graduating class since the members of the Class of 50 had been graduated by June 1950. More details about the Clemson Arnold Air Society are listed in the appendices.

During 1953 the AFROTC staff at Clemson now consisted of Col Lloyd L Hull, the Commander and a staff of 18 personnel. The Arnold Air Society had 15 members and was a relatively young organization on the Campus but it continued to grow.

During 1955 there was some reluctance of the Veterans that were entering College to participate in the routine cadet exercises and this led Clemson College to an official decree on July 18, 1955 that brought an end to military training for all students and this in turn permitted the College to become Co-Educational. Undergraduate women were admitted during the second semester of the academic year of 1954-1955. During 1958 and after the sudden death of Dr. Robert F. Poole on June 6, 1958, who had recommended that Clemson create a graduate school, the trustees chose Robert C. Edwards, a Clemson alumnus and Textile executive who earlier had been the vice president for development in 1956 at Clemson, to become the president of Clemson. The science programs at Clemson were being upgraded and the first doctoral degree was awarded in 1958. With less emphasis on the military and with the addition of co-education, Clemson was now in a position to become a University, however this did not occur until March 11, 1964.

Now let us return to the year 1946 and the third Clemson Aero Club's development. A new breed of students entered Clemson that year and many were interested in a career in aviation but courses in aviation were not offered at Clemson. A number of students were interested in the Air ROTC program that was now being offered there at Clemson in the fall of 1946. But in the beginning the curriculum followed the Army ROTC program. The Air Force AFROTC students attended an Army Class three hours a week for thirty-two weeks. The cadets were told that during the third and fourth years that the Air ROTC program would provide courses that featured a specific occupational specialty as adminstration, aircraft maintenance, communications, etc. Cadets were required at the end of their third year to attend an Air Force Summer Camp at an Air Force School that offered specific training in their specialty that they were planning to continue after their commissioning as a 2nd Lt. USAF Reserve. During their senior year they were paid $27.00 per month for participating in the AFROTC program. There was no required mandatory active duty commitment at the time of graduation but the reserve officer was required to remain in reserve for 5 years and be available to enter active service if required like the Army ROTC requirement. Regular Air Force Commissions were offered to Distinguished Military Graduates (DSM) who had been recommended by their Professors of Military Science and Tactics. The regular officers were required to enter active duty immediately upon graduation and serve for a period of 4 years.

Since there were no aviation courses being offered at Clemson it was natural that there might be a demand for the development of a flying club because many of the new students and some of the returning veterans were now interested in continuing their previous military careers now as officers in the USAF and careers in Commercial Aviation were now developing also.

One of these new students was Robert R. Russell . He entered Clemson as a freshman in September of 1946 and being an aircraft model builder, having built his first model airplane in 1949, decided that he wanted to learn to fly a real aircraft. During his sophomore year he, with the permission of his father, began working on his student pilots license at the local Clemson airport that had just been in operation for a short while. Faced with the cost of learning to fly and going to College he decided that some kind of Flying Club should be available for Clemson students to learn to fly at a more economical rate. So in the fall of 1947 he organized a special meeting to be held in his dorm room asking those students who were interested in forming a Flying Club and requesting that those interested should attend this kick off meeting. There were between 20 to 30 people who attended the initial meeting that night and as a result there were finally 21 students and WWII veterans that agreed to form a Club. From that initial meeting enough interest had been created and by the following spring the 21 club members had agreed to purchase an aircraft, a J-3 Piper Cub, if one could be found at a reasonable price. By this time Bob had been well into his flight training there at the Clemson-Central Airport, his instructor being Ernest E. King, who later soloed him at the Anderson Airport, Anderson, SC. Bob was now a certified pilot so he searched around and found a J-3 Piper Cub at the Greenville Airport that was for sale. Each member bought a share of stock for about $38.00 to be used to purchase the aircraft. Bob went to Greenville, purchased the aircraft, and the flew it to the Clemson-Central Airport and now the Club could began its flying operation. .

Mickey Dorsey was the first President, Robert R. Russell was the Vice President, M.M. Burley, Secretary, J.M. Jackson, Director, J.F. Craig, Treasurer, and J.D. Hogan, Business Manager. The member dues were $2.00 per month to pay for the maintenance and hangar rental space cost. Each member paid for the fuel used in his flight instruction and the instructor's fee was $7.00 per instruction. The author was one of the charter members of the 1948 Aero Club and already had a private pilot license.

The first flight instructor was James M. Jackson, a WWII pilot. Jackson began his flight training in his hometown of Rock Hill, SC as a member of the government sponsored Civilian Pilot Training Program (CPTP) in 1941.

They started out in a competitive night time ground school and the top 10 students were selected to receive 35 hours of flight training in a Piper J-3 Cub. After completion of this phase and receiving his private pilots license he was placed in the US Army Air Corp reserves. He was then assigned for secondary training in Spartanburg, SC using the Waco UPF-7 for this phase. The Stinson Reliant was used for their navigation training and it also was group training aircraft in that there were 3 students and an instructor for this phase. The Stinson flights were long navigation flights and later there were individual short navigation flights in a Taylorcraft of 4 hours duration (2 hours out and 2 hours back). He then went to flight instructors school at Griffin, Ga. and graduated as a flight instructor for the CPTP. He taught there for a while and after serving for the required time he was allowed to resign his position and to enter the US Army Air Corp as an aviation cadet.

After completion of the Aviation Cadet program he was graduated as a 2nd Lt and was awarded his pilot wings. Sometime later he was released from service and enrolled at Clemson A & M College in 1946 for his college education and became an instructor for the new Aero Club in 1948. Later in 1950 he became the president.

On May 25, 1949 a Certificate of Incorporation for the Clemson Aero Club was issued and was signed by W.P. Blackwell, Secretary of State with the following people as officers: J.M. Jackson, President, J.R. Rostron, Director, C.D. Foster, Business Manager, J.F. Craig, Treasurer, and W.R. Hale, Secretary. The certificate listed the Headquarters as Clemson, SC and state the purpose of the organization is "To provide flying time for Clemson men at a minimum cost" and "they desire to be incorporated in perpetuity"

Their flying operations were to be conducted from the new Clemson-Central Airport that had just been completed and was ready for operation. This airport was developed by the Meldeau Flying Service that had moved their flying operations from the Seneca, SC airport.

The Meldeau Flying Service was operated by Leonard H. Meldeau, a much decorated WWII fighter pilot and a former Seneca citizen. Meldeau grew up in Seneca, SC and as a youngster was a builder of model airplanes and he always wanted to fly. He was a charter member of the Seneca Model Airplane Club which was started around 1935. There were six other members: Douglas Bradberry, who graduated from Clemson and later became a Professor of Engineering Drawing at Clemson after his military service, John J. Harper, who graduated from Clemson and also Georgia Tech, and who later became a Professor of Aeronautical Engineering at Tech, Otha H Vaughan Sr., Harold Abbot, Clyde Lother, and Marvin Barron. The author also built model airplanes with his father during this time and enjoyed being associated with these people and their club as they were excellent role models.

In 1939 after graduating from high school Meldeau built an aircraft which he flew and in early 1941 he decided to join the U.S. Army Air Corps but not having the required two years of College education required for training as an aviation cadet. Since he was not eligible he volunteered for service as an aviation cadet in the Royal Canadian Air

Force . While there in training he was first in his flying class, graduating with highest honors for excellence in his studies and flying. He was sent to England and assigned to the 421Canadian Squadron of the Royal Air Force for combat. He flew 50 combat sorties with that group and he later in September 1942 joined the U.S. Army Air Corps, 31st Fighter Group, 309th Fighter Squadron and while there he distinguished himself in flying some 125 more combat missions over North Africa, Italy, Sicily, and Malta. During those mission he shot down four German Me 109 aircraft, damaging nine aircraft, and two probably destroyed. He was shot down as he was taking off for a combat mission, severely injured, and then sent home to recuperate. After he was released from service in 1945 he returned to Seneca, SC, his home town, and started his flying service operation at the small airfield located east of Seneca and north of the Utica Textile Mill village. The field was owned by Bert Tribble of Seneca.

Aviation became a part of Seneca's history when shortly after WW I three individuals decided that Seneca needed an airport. Those individuals were Bert Tribble, Major Linsay, and Oscar Wigington who were interested in the new field of transportation. Bert Tribble was the most active of these individuals as he provided the land for the airport. During the early 20's the barnstormers would come the Seneca airport and offer sight-seeing rides. The aircraft were the surplus WWI Jenny's that these pilots flew their passengers in. As late as 1935 this airport was listed in the 1935 Air Pilots Register as an auxiliary field for use. On one ocasion sometime around 1935 a pilot who was flying an Oil Company's aircraft from Atlanta to Charlotte, NC was running low on fuel so he landed at the Seneca airport. My father and I saw him land and we went to the airport. After talking with the pilot we went to Seneca to get him some five gallons of fuel. We then loaded the fuel in the aircraft using a chamois skin to filter out any water that might have been in the fuel container. After the fueling was completed the pilot offered to give me and my father a ride. This was my first ride in a Stinson Reliant aircraft as I sat in my fathers lap and the pilot let me fly the aircraft. I was hooked at this time to be a pilot of some type of aircraft in the future.

I always wanted to be a pilot as long as I could remember and my father, although not a pilot, encouraged me to learn about aviation. In July, 1944 I joined the Civil Air Patrol Squadron of Anderson, SC as a cadet so that I might learn about aviation and the military as I wanted to be a military pilot in the future.

Meldeau was my second role model as well as my father and so in October 1945 I began taking flying lessons from Meldeau in an Aeronca L-3 that he had purchased as surplus from the government. After a few hours of instruction in the Aeronca and then in the Piper J-3 Cub, I was ready to solo. I soloed on a Sunday in the afternoon of April 26, 1946 and on Monday I was graduated from Seneca High School. I was the first of Meldeau's students to solo from the Seneca airport. After I had soloed, a number of Meldeau's students also received instruction and soloed from the airport later that year.

Meldeau operated the flying service for the rest of that year and then he and Joe Salvadore, an executive with the LeTourneau Corporation, and with some other stock holders, formed a corporation "The Meldeau Flying Service" and decided to relocate his present flying operation at a new airport that they would design and build near the Clemson A & M College campus. Meldeau was the President and Salvadore was the major stock holder. Meldeau designed the proposed airport to have two runways, one that was orientated in a North and South direction and the other roughtly in an East and West direction and located about 2 miles from the campus. He enlisted his brother-in-law Jim Smith to help him in the the construction of the hangar and the grading of the runways for the airport. He was the first Piper Aircraft Distributor in South Carolina and offered Piper aircraft for sale, flying lessons for the general public, and a government sponsored Flying Educational Program under the GI Bill. The GI Bill program allowed the veteran to enroll in a college to work toward a degree or the veteran could take vocational education courses if he desired. Meldeau Flying Service offered a private pilot training course so that a veteran could to get an entry level job in the aviation fields that were just beginning to develop. There were about 11 to 15 veterans enrolled in the initial program. A few of these veterans that took the GI Bill aviation courses went on to get their commercial and instructors tickets and were later employed by other flight training operators. The aircraft that were available for training at first were a Piper J-3 Cub, a PT-22 Ryan.(a former U.S. Army Air Corps primary trainer), and an Aeronca L-3 (a former U.S. Army observation aircraft). Later a Piper AE-1, a Super Cruiser that had been used during WWII as an ambulance aircraft, was added to the fleet of aircraft and a Taylorcraft was purchased. The Cub was used mostly for general aviation orientation rides, aircraft rentals to individuals, and training operations. The PT-22 was used to give orientation rides for the general public and was at one time a rental aircraft unil it was crashed around late 1947. The Stearman PT-17, a former primary Army Air Corp trainer, Fairchild PT-23, a former Army Air Corps primary trainer, and a Fairchild PT-26, a former Army Air Corps primary trainer were owned by Jack Gibson. The Clemson Aero Club owned their Piper Cub, N41282 and they acquired another Cub, N41026. Later a Taylorcraft, N39107 was also purchased sometime in the late 1949 or early 1950.

Meldeau left the Meldeau Flying Service operation sometime in late 1947 and went to Norfork, Va. to work for Glenrock Flying Service. He later was employed by the Skyline Flying Service in Asheville, NC. where he was a flight instructor for a while. He decided to return to Greenville, SC where he was employed by Southern Airways as a flight instructor, teaching fledgling student pilots at Bob Jones University. In 1949 Professor John J. Harper of Georgia Tech and a close friend of Meldeau suggested that he go to college to study for an Aeronautical Engineering Degree and so he enrolled at Georgia Tech that year to work toward his degree. While enrolled at Georgia Tech he joined the Georgia Air National Guard (GANG) and flew P-80's and F-84's. He flew with the GANG until they were called to active service during the Korean War where he reported in the fall of 1950 to George AFB, California and he instructed the new USAF pilot in combat techniques. He remained on active duty with the USAF after the Korean War was over, and after serving in a number of air force locations in the states and in Germany he retired in 1969 a Lt Col. and a Command Pilot. After his retirement he worked as a manufacture's representative for a while and then started his own business of providing specialized explosive hardware to the Aerospace companies and others and then to NASA in support of the early Space Program at the Kennedy Space Center. He died in 1995 after a long battle with Cancer.

During the period of 1947 through 1950 the airport was operated by Ernest E. King, a designated FAA Flight examiner and a former US Army Air Corps B-17 Flight Engineer. During 1947 through 1949 William F. Wilson was an instructor for the Meldeau Flying Service. William was a former B-24 pilot during WW II and had come to Clemson to work toward a degree in Engineering. He graduated BS Civil Engineering Class '49 and went to work with a Civil Engineering firm in SC. Robert Parker was the A & E mechanic for the airport and also provide maintenance service to owners of their private aircraft that were based at the airport.

Mr. King initially was employed by the airport manager at the LeTourneau Airport at Toccoa, Ga. where he was an flight instructor during a period of time in 1946. He later accepted a job in 1947 with the Clemson-Central Airport and commuted from Toccoa in his Ercoupe. He later lived in the old farm house that was on the property and continued to operate the airport until about 1953. About this time the corporation was contacted by a Mr. William Martin, a chicken farmer that wanted to get into the business of raising a large number of chickens for sale and inquired if the property was for sale. At that time the airport business was not too good and so the property was sold. The deal was closed and the Meldeau Flying Service ceased to operate but the Flying Club was allowed to operate their aircraft until another airport could be found for their operation.

The flying club by late 1950 had about 38 members and had two Piper J-3 Cubs, N41282 and N41026 and a Taylorcraft, N39107. By 1951 there were about 30 Club members. The club operated under a set of By-Laws as follows:

Clemson Flying Club By-Laws

1. Nominations from the floor must be called for by the presiding officer at the election of members. Members must be enrolled only after a favorable vote of 2/3 of the active membership and their purchase of (1/3 cash) one share of stock amounting to $40.00. Membership will be open to all classes .
2. There will be no initiation whatsoever.
3. Besides the appointed faculty advisor, there will be a club president, vice president, secretary, and treasurer. A vote of 2/3 of the active membership will be required for any person to hold any one of the offices. The president will be a qualified private pilot.
4. The vice president takes over in the cases of absence of the president.
5. Meetings will be held the 1st and 3rd Thursday of each month at 6:30 o'clock, room 100, Riggs Hall, or unless otherwise specified.
6. Any club member, living on the campus , who is absent from the meeting, without a legitimate cause will be fined 50 cents.
7. An appointment book will be kept in the guard room at all times and persons desiring to fly will put their names, and the time they desire to fly in the book.
8. The maximum affirmative vote or the election of a member must not be greater than 3/4.
9. No one may fly the club airplane except the regular instructor and/or a shareholder who has been properly checked out in the aircraft.
10. There will be a charge of $2.00 per hour of flying time for the use of the aircraft.
11. There will be a fee of $2.00 per month for all members in the club to cover insurance, depreciation, and inspections of the aircraft.
12. All dues , fees, and bills are due on the 10th day of each month. Any member who fails to pay his bill at the designated period will be grounded until the stated amount is paid.
13. No dual instruction will be logged except by the instructor or by a member holding an instructors certificate and approved by the instructor or by his aid.
14. No flying will be done until the $40.00 share of stock is paid in full.
15. A log book will be kept at all times in the aircraft and each member will register his name, time in, time out, and the date the aircraft was flown immediately after the flight.
16. A member may be expelled (by a 3/4 vote) on proof of dishonesty or disgraceful act in connection with the club business or club activity. His share of stock is automatically forfeited.
17. No share of stock may be sold to anyone unless the person has been approved for membership in the club by regular ballot.
18. No person may fly the aircraft until properly checked out by an instructor.
19. Any repair work done by any members will be approved by the president and the faculty advisor under the supervision of Carolina Aero Service.
20. Any malfunction of the aircraft will be written in the aircraft logbook and reported to the Carolina Aero Service.
21. The aircraft will never be landed on a non-operational airport except in case of emergency. Any member guilty of violating this rule will forfeit his stock in the club.
22. The aircraft will never be flown lower than the altitude prescribed in the Civil Air Regulations. All flights will be accordance with the CAA,. Any member violating any of these rules will forfeit their share in the club.

23. Only qualified members may take the plane home or on a cross country trip. Trips in the summer time will be approved by the faculty member or by the club president.
24. A fee of $1.00 per day will be charged to a person keeping the aircraft at home in the summer besides the $2.00 per hour flying time and gas and oil.
25. Any member who damages the aircraft during a flight or handling will be reported immediately to the faculty member or the club president.
26. Any situation arising other than the above stated will be discussed by the president and faculty advisor and may be voted on if deemed necessary.

Mickey Dorsey attended Clemson during WW II and had played football and was in great physical shape. His room mate James Dickey there at Clemson had joined the Army Air Corps and told Dorsey how great it was so he decided to join the Army Air Corps. However he was born with only a thumb and one finger on his left hand so he was rejected by the Army Air Corps. He then tried to enlist in the Charleston Army Air Base but was turned down again. He then tried to enlist in the Marines and was rejected again. He next tried to enlist in the Navy but again he was denied. Still determined he was finally accepted in the Army for Limited Service--meaning you cannot be sent to combat but you will remain stateside and train troops. He was then sent to Ft. Jackson, Ft. Bragg, NC and finally to Camp Blanding, Fl for 17 weeks basic training. However he was one of three persons (In a Company of 200) to complete 100 % of the physical and endurance tests). He also had the highest score on the rifle range for the entire battalion of 1000 men. This resulted in him being assigned to the 71st Infantry Division. This division began its action with General George S. Patton's Third Army in France. There he battled across Germany and into Austria. During his service he was in the first group to liberate Gunskirchen Lager Concentration Camp near Lambach, Austria. He was awarded a large Jewish medal from the Jewish Community Relations in California and later was awarded at the Jewish Heritage Center in New York, the Jewish Soldiers Medal by the Nation of Israel, the only time that that medal has been awarded to anyone other that Jewish Soldiers. In 2006 he was given a certificate and medal and was created an honorary member of the "Austrian War Veterans Association," his former enemies.

Mickey Dorsey the first president of the Clemson Aero Club, a WW II veteran, a private pilot, and also an aircraft owner of a Piper L-4 Artillary Spotter Aircraft that was declared surplus after WW II.

He remembered some events that occurred while he was president. One particular event was that since one of the airport's runways was red clay and sloped downhill, it seems that every time there was a rain shower that the rivulets of water would create small crevices in the runway making takeoff and landings exciting. After some brainstorming to solve the problem the club decided that they would make the president of Clemson College, Dr. Robert F. Poole an honorary member of the club. They made up a special "Clemson Aero Club Honorary Member" Card. It was presented to him in his office and he was invited for a ride in the club's aircraft to view Memorial Stadium and the Campus from the air. As Dorsey was taxing the aircraft, he kept apologizing for the rough field and asked Dr Poole for his advice on how it might be made smoother. They flew around for about one hour that day. After landing Dr. Poole said he really enjoyed the trip. The next day a Clemson road grader was at the airport and the runway had been scraped smooth. Seems that after each rain that the runway always appeared to have been scraped smooth.

Another incident that Dorsey recalled was that since the Club's aircraft needed to have an annual inspection he and Joe Gault decided that he would fly his aircraft and Gault would fly the club's aircraft and Dorsey would fly him back to Clemson. So Joe took off and Dorsey said that he would follow him and fly beside him to let him know that he was there with him as they flew to the Greenville Airport. Since Dorsey wanted to be a fighter pilot he decided that he would shadow the other aircraft by flying behind, above, or below Joe's aircraft. Every time Joe would turn or descend Dorsey would remain out of sight and after landing Joe said, "I thought you were to fly with me but I never saw you." Dorsey replied, "I was there."

Another exciting event occurred while he was teaching Randy Hale to fly. Dorsey was in the front seat and Hale was in the rear seat that the student pilot occupied. Dorsey had instructed him in 90 degree turns and 360 degree turns and now it was time to learn how to do power on and power off stalls. After one or two stalls Dorsey said he felt something hit his right shoulder and Randy had become sick and indicated he was about to lose his cookies so Dorsey opened the side door and headed down to land but Randy didn't make it and vomit covered the side of the aircraft and the elevator. Naturally he was required to clean up his mess. Later he progressed nicely and later became a pretty good pilot.

During April while flying the Taylorcraft T. L Cox and his passenger, B.K. Faulkenberry had to make an emergency landing in a pasture near Pendleton when their engine failed. The landing was successful but as the aircraft during the landing came to a slow stop, one of the landing gear hit a depression, damaging it and the aircraft nosed over damaging the nose cowling where the carburetor air scoop was located. The aircraft was repaired for a ferry flight and was repaired later and continued to be available for club members for flight.

During May 1950 eight aero club members and two faculty advisors attended the National Inter-Collegiate Air Meet held at Stephens College, Columbia, MO. Attending were J. R. Carter, W.R. Hale, T. L. Cox, B. K. Faulkenbery, T. R. Hawkins, B.J. Mullin, J.B. Rast, E.H. Shelley, Prof. J.B. Long, and J.P. Rostron. Stephen College acted as hostesses and provided the program for the weekend. The Civil Aeronautics Administration furnished judges for the various competition events. The Clemson pilots entered in several events which included flying a three leg closed course. A map was provided to each crew. The purpose of the event was to estimate your time of flight around the course and the winner was decided by who completed the course in their estimated time after flying over two specfic markers on the course. The Clemson pilots didn't win that one. There was also a bomb dropping contest in which each crew was given 5 paper sack bombs to drop from a specfic altitude these bombs at a target on the end of the runway. The Clemson pilot and his partner came close. There was also a spot landing contest in which the aircraft must be landed as close as possible to a line on the runway. They gave the pilot three tries and best of the three was your score. The aircraft that won this contest was a military surplus trainer, a BT-13 from Ken State. This aircraft performed much better in the wind that the small tail dragger aircraft that the Clemson pilots were flying since it was a heavier aircraft.

Euel H. Shelley who was a member during 1950 -1951 recalled that William Hale, also a member of the Clemson Aero Club, and someone else flew a surplus Army Air Corps Vultee BT-13 aircraft from Greenville, SC to the meeting in Columbia, MO. Euel drove his automobile and took 2 or 3 other club members with him. William Piper, the owner of Piper Aircraft, attended the banquet that was held at the meeting. The weather Euel recalled was really windy. It was so bad that as the little aircraft would land personnel would go out to the runway and hold on to the struts to keep the aircraft from becoming airborne again. He had a private license before he joined the Aero Club. He was drafted and entered the Army as enlisted in 1951. He went through basic training and leadership school at Ft Jackson and served there for a short while before being assigned to Company I of the 17th Infantry Division and saw service during the Korean War. He was awarded the Korean Service medal with bronze star, the United Nations Service Award, and the Combat Infantryman's Badge Award. He was released from active duty with the rank of Sergeant. He returned to Clemson and entered Clemson again to complete the requirements for a VAE degree, which he received in 1954. While attending Clemson he purchased a Cessna C-140 aircraft and flew it out of the Anderson Airport where he rented hangar space from

Ben Johnson's flying operation. Recollecting his aero club experience before going into service he remembers flying the Aero club's Taylorcaft and Piper Cub as well as an Ercoupe aircraft that was there at the Clemson-Central Airport. He recalled the J-3 Piper Cub being shared among the members in the summer months. William Hale, a club member, flew the J-3 to Nichols, SC for him to use it for a while and later he flew it to Latta, SC so that Tommy Squires, another club members could use it . Later after leaving the Clemson area he had a number of interesting experiences while flying the C-140 aircraft. Sometimes he would fly it to the outer banks near Morehead City, NC and land on the beach there to do a little fishing. He later joined the Civil Air Patrol as a Senior member and received his commission as a Captain. During one of his many missions flying for the CAP his engine cut out but he got it started again and continued to fly the mission.

During 1952 thru 1953 the President was Randy Hawkins, N.R. Tice was VP, H.R. Hogg, Secretary, C.A. Whitaker, Treasurer. The club had about 23 members and the club had a Taylorcraft, N39107 and a Piper J-3 Cub.N41282.

The club sold the Taylorcraft, N39107, to Randy Hawkins, a club member also during 1953. Randy tells of an incident that happened to him and a friend Jack Spillers while Randy was still a member of the flying club and his story is as follows:

Since Randy and his friend were already "experienced pilots" they were ready to make the trip to New Orleans, La. to participate in a Senior Platoon Demonstration rather than ride down there in the Clemson Bus with the other guys. Aside from lacking the proper Aeronautical Charts and a two way radio for two way communication and adequate cross country flying experience they decided that they would fly to Athens, Georgia and then proceed south to the coast. From there they could fly west until they got to New Orleans, weather permitting. They took off that morning and after the traditional toilet paper roll toss while passing over the campus they proceeded toward Athens, Ga. After flying for a while, there appeared Athens, Ga. After they landed it was Athens Airport after all. However after they had left the Athens Airport and were airborne for about 30 minutes and with slightly deteriorating weather conditions with light rain and lowering of the clouds and visibility the aircraft began to experience a slight vibration. Randy decided to land so they began to look for a suitable site and they found an open field that looked ok. He decided to land in the tracks of a road path going across the field. After landing they discovered that the rest of the field contained many stumps covered by grass. Inspection of the propeller revealed that some of the canvas under the prop tip sheath had come loose. A little airplane glue did the trick and after waiting for the glue to dry they decided to take off again, heading south in the rain. A few minutes later however the aircraft began to vibrate violently and would not maintain altitude. Randy headed for the next best thing, a small cow pasture. After almost clipping the top of a Trail Ways Bus, bouncing over a three foot ditch, and finally ground looping the aircraft, they came to rest in a farmer's back yard. You can imagine the the farmer's wife's surprise when she responded to Randy's knock on her back door and saw an airplane close to her doorsteps.

To make a long story short, Ramsey and Jack spent the next two days hitch hiking to New Orleans. They arrived in time to participate in all the Senior Platoon's activities and spent a bewildering night on Bourbon Street. The bus trip back to Clemson was somewhat humbling for the two aviators but they were grateful not to have suffered any greater consequences. An interesting note to this story is:

Randy says that Jack Spillers (more details on Jack Spillers later) sent a post card to his mother a day after they had their forced landing and it plainly said:

Dear Mom,
We Crashed
Love,
Jack

What a sense of humor!!!!

Several week later, after purchasing another used propeller, the two drove to Georgia and retrieved the aircraft. When they arrived with the other prop they found that a bull had used the lower cowl of the aircraft as a back scratcher and had mashed the cowling up against the engine. Randy says they did a little dent repair before they took off. They also bought gas from the country gas station that the plane was parked behind.

Randy was graduated in 1953, entered the Army as a 2nd Lt, and took his Taylorcraft with him. He later traded the Taylorcraft for a Stinson Voyager that he kept until he was released from active duty. He used his airplane in his dairy farming operation and when Clemson began to offer bull semen to the dairy farmers he flew the semen, sealed and iced in tin cans to the dairy farmers in the state to breed their cows. He developed a medical problem in 1983 and stopped flying but he cherishes the experiences and the memories as a club member.

Another Clemson graduate W. C. Mooreman, during his senior year in 1950, decided to join the flying club and get his pilots license. He finally saved enough money so to join the Flying Club, later becoming a member and buying his share. He stated that the the flying club operation was nice to deal with since flying was cheap, relaxing, and an easy way to travel on weekends after he had received his student pilot license. He flew his solo cross country from the Clemson-Central airport, to Henderson-Asheville Airport, to Spartanburg, and then back to Clemson. After his solo cross country flight he flew a number of sight seeing flights to the mountains. One day he was flying the Taylorcraft down to Aiken to attend a weekend retreat however a problem developed during the flight. While he was flying and looking out the side window searching for the location of the retreat camp his glasses blew off and out of the aircraft. Needless to say he was still able to find the airport and land safely. In another instance he was flying to Winston Salem to see a girlfriend at the College there and no problems on the flight. The next morning he checked the weather before leaving Winston Salem and the weather report was good. So after leaving the Winston Salem area and climbing to his requested altitude he noticed that he now had a ground fog that obscured the ground and the river valley below. The event was really no problem as flying conditions and visibility were great. Flying by his compass he flew on in the direction of Florence his next stop. As he got near Florence he now was flying between layers of clouds but as he got nearer to Florence he was able to see the airport and land safety. Since he did not have a private license he decided to start working on the license and finally got his private ticket some time later. He also rented and flew an Ercoupe that was at the Clemson-Central Airport while he a member of the Clemson Flying Club. The flying club was great fun. He stated he enjoyed the "hangar talk, and cost of flying was so cheap in the 50's".

Tom Squires, as a member of the 48 Aero Club, received his private pilots license in July 31, 1948. He remembered that on one of his cross country flights from Rock Hill, SC, Lancaster, SC, and then to Marion, SC he had no radio in the aircraft which was common at that time. So navigation was by time, compass, visual observations, and a sectional chart. On his flight he encountered a 15 knot crosswind which blew him off course so that he flew over a small town with a water tank that had Jefferson, SC painted on it. Knowing this he found himself on the map and using his compass continued his flight to his destination. He was graduated from Clemson in January 1952 with a degree in Animal Husbandry and his 2nd Lt. commission from Army ROTC. He entered the Army in April 1952 and got married in August 1952. He went to Korea in December 1952 and served as a Platoon Leader, Infantry, and also made a number of flights in a Cessna Bird Dog aircraft not as a pilot but directing ground fire as an Artillery Observer. He was released from active duty, started his family, and entered the work force. He is now retired from the U S Department of Agriculture and is living on his farm near Latta, SC. He is one of the oldest Auctioneers in the state of SC, and he is still active.

Fred Cone, also an Aero Club member, provided me with a couple of stories and here they are:

Fred was in the Army during WWI and after his release he worked for a while at a Lumber Company and began his flight training under the GI Bill that was offered to returning veterans. His instructor was a former B-17pilot, Charles Ward and Fred soled at the local airport at Elmore, Al. He later recieved his Private certificate at Dannely Field, Montgomery, Al in 1946. He enrolled at Clemson in 1947 using his remaining GI Bill funds and began playing football for Clemson's Frank Howard who had invited him to come to Clemson.

After Fred joined the Clemson Aero Club he and his brother Mickey decided that they would fly the Club's Piper Cub to their home in Coosada, Al. They started out and due to a pretty strong head wind they finally arrived at Gainsville, Ga, after about 2 hours of flight.

There they refueled for the next destination headed toward Atlanta and they finally arrived there where they again refueled. Next they flew to Auburn's airport and there they landed again for fuel. The next stop was to be the airport at Elmore City, Al airport. Since they had not told any of the family that they were coming home they decided rather than walking the 4 miles to their mother's house that they would fly over there to get their attention to come to the airport to pick them up. As it was David, the handyman, was out mowing the yard and so Fred as he was flying at about 500 feet and as they passed over David he throttled back the engine and yelled out to David "Get someone to come to the airport to pick us up". David was quite shocked, dropped his lawnmower, and he ran into the house and shouted "Mrs. Maggie, I just talked to Fred!! He just flew over in a yellow airplane and said to send someone to the airport to pick them up." Fred's mother said "Are you crazy?" David then said "I swear to God Mr. Fred flew over and said 'Come to the airport he need a ride!'" Needless to say sometimes we can communicate without a cell phone (they weren't available in 1948!).

Another story that Fred recalled was that he and Don Wade decided one day that they would fly over an area where the Clemson and Carolina teams were playing in a baseball tournemant and they would drop out a chicken. They came flying over at about 100 feet, released the chicken, and scored a direct hit on second base. Needless to say they were immediately grounded, were told to drop their membership, and they didn't fly the club's aircraft anymore. Prior to this event Fred remembers that he and Don would fly in different aircrafts and chase each other in the area of the "Bottom Lands of the Seneca River" there south of the campus.

There was also another story reported to the author that a Clemson pilot and a friend flew down to Columbia in their own airplane and proceeded to pitch out a chicken as they flew over the Carolina Football stadium during the half time of one of the Big Thursday Clemson-Carolina football games.

Also another member of the Clemson Aero Club recalled that two club members, a Senior and a Freshman, decided they would fly down to the Isle of Palms to see a girlfriend of the Senior. On the way down as they approached the Cooper River Bridge they decided that they would fly under it. So after a discussion about WWII movies they had seen which showed a B-25 flying under a bridge they decided to do it and as they passed under the bridge they waved at the passengers that were on a Gray Line Tour who were also passing under the bridge. Needless to say they later landed on the beach and had a very enjoyable weekend with the Senior's girlfriend and her family there at the Isle of Palms. The trip back to Clemson was uneventful.

Jim Woodfin was a member of the flying club in the 50's also and after his solo and sometime later he decided that he would fly to Winthrop as he had a girlfriend there. He left the Clemson Airport a little later than he had planned so it began to get dark as he was flying the Club's Piper Cub so he managed to locate a field near Rock Hill and after flying around for a while some people came out and turned on their car lights and he landed successfully. Returning to Clemson the next morning presented a problem since it is easy to land in a small field but the aircraft requires more runway for a take off however this was solved by contacting a Highway Patrol officer to clear the road near the field, after he taxied the aircraft to the road , and this new runway allowed him to take off. The flight back to Clemson was no problem. More details on Jim are in Chapter 13.

During 1950 the Aero Club's Piper Cub was also used by Professor Joe Rostron as an aerial photographic platform a number of times. He mounted a camera on the right wing strut and took aerial photographs of Clemson, SC, Pendleton, SC, and other targets of interest. The photographs of the towns were then used by his students who were taking a course in Photogrammery that was being taught there in the Civil Engineering Department. Detailed maps were produced of each town by the students. Joe was also a Flight Instructor for the flying club during this time as well as a faculty advisor.

There was a second airplane, a small biplane called the Knight Twister, that was built by a Clemson student John J Weeks using the school's woodshop and machine shop during the 50's. He provided me with this information about his building an airplane while he attended Clemson A & M College. He enrolled in Clemson as a freshman in 1947 majoring in the field of Mechanical Engineering. He did not join the Clemson Flying Club that had begun its operation in 1948 since he already had a pilot's license. He decided during the start of his senior year in 1952 to build himself an aircraft in the Industrial Engineering Workshop where the first airplane that was built at the college was flown in 1929. This was the second aircraft to be built in the woodshop. While taking an aeronautical course he proposed as an engineering project to Prof J. Marshall, the building of an aircraft wing. He selected the Knight Twister, a small acrobatic type biplane. Since this was not a kit aircraft as we know them today, all he had was a set of blueprint drawings from which he used to build the wing and later the whole aircraft. He didn't have very much money so he began to look around for materials to build the wing. Since the wing design for the Twister was of a tapered design he was able to find a Taylorcraft wing that was not damaged but it was of a straight wing design. The wing spars were in excellent shape so he was able to use them however the fact that the wing to be built required tapered configuration he was forced to make his own ribs since each rib airfoil was of a different size and shape and length. The top wing after it was assembled was covered with plywood, and the bottom wing construction was also completed in very short order. He next obtained a fuselage from a damaged Aeronca Chief and after much reshaping (cutting and welding of it)in the machine shop during the spring and summer of 1950 a fuselage began to take shape. By using as much of the original tubing he was able to modify the Aeronca fuselage to give him the Knight Twister basic configuration he needed. The fuselage was built up of a series of formers and he salvaged a number of wooden stringers from a Stearman aircraft to get the required configuration of the fuselage for the Knight Twister. After the fuselage was completed he was able to find Cessna spring steel landing gear and mated it to the existing fuselage. Aircraft instruments were removed from another damaged aircraft, refurbished, and added to his aircraft. The main wheels with their tires and brakes and the tail wheel were from the Aeronca. The horizontal stabilizer with its elevators and the vertical stabilizer and its rudder were constructed of wood and were completed by also covering them with fabric in his dorm room during that summer. The aircraft was assembled in the machine shop later that summer using all of the required parts to make it a complete aircraft. The first flight of his aircraft would have to wait till later in his career. He was graduated with a degree in Mechanical Engineering during the summer school session, receiving his USAF 2nd Lt. commission, and a week later entered active duty with the USAF at Wright Patterson Air Force Base , Dayton, Ohio reporting for duty as an Aircraft Maintenance Officer. However before leaving for his active duty he had the aircraft transported to Charlotte, NC and put in storage. It remained there till the summer of 1954. After his return from active duty in summer of 1954 he began flying his aircraft for the next few years. In 1956 he sold the aircraft to a pilot in Akron, Ohio and purchased a Culver Cadet. While working for the Trane Company he was relocated to Hawaii and had a very long successful business career in air conditioning and sales for the Trane Company as their representative. He has owned a number of other types of aircraft over the years while living in Hawiaii, Florida, and other places before moving back to NC where he is now semi-retired, devoting his time to Real Estate Development and Educational Seminars.

So aviation has had quite an influence on his career.

In 2003 the little Knight Twister, after being owned by a number of people, was located by the author in Kendrick, Idaho and was in the process of being restored and modified by Fred J. Ring and John Rowden. Ring died before it was completely refurbished. His family donated it to the "Wings of Hope" organization of Chesterfield, MO and according to the operators there the engine and other parts were sold and that the aircraft no longer exists. It would have been nice to have obtained this aircraft for display at the S.C. State Museum where "Little 372" is now displayed.

The Flying Club continued to operate from the Clemson-Central airport as it was called. Sometime in 1953 the Piper Cub J-3, N41282 was totaled in a crash. The student pilot was injured but not critically. He suffering a head wound. The pilot was C. Emory Bishop, who was flying from the Anderson Airport. The crash was as a result of him seeing some kind of obstruction on the east to west runway and so he then decided to land on the north to south runway. He came in high and decided to do a go-a-round and somehow lost control during the go-a round and crashed. He suffered a severe head wound as well as multiple bruises. The aircraft was completely destroyed. Bishop had entered Clemson in 1952 and joined the flying club. Since his father served in the Balloon Corps during WWI and his uncle was a fighter pilot during WWII Bishop had always wanted to be a pilot as a child. He served as Vice President of the flying club during 1952. After soloing he made a number of cross country flights as a club member. He was drafted near the end of his sophomore year and entered military service. After his release from service he returned to Clemson and continued his education under the GI Bill program studying Animal Husbandry. He was awarded his degree. Since he was still eligible for more schooling under the GI Bill he decided to continue his flying training. He was later awarded his private, commercial, and multi-engine ratings. He continued his love of flying with his own Cessna 172 and later flying a rental Cessna 182 on occasion. He also flew a Beech Baron B55 and a Beech King Air. He decided to retire from flying in 1996. He retired from the S.C. State Farm Bureau after 28 years of service to that organization. In retirement, he still operates a farm and has other business interests. His association with the flying club also played an important part in his later career.

During the summer of 1955 Clemson as a Land Grant College ended its association with the Military ROTC program. The Land Grant program required that all students participate during their first two years at school in a military training program and during their last two years they could elect to continue to participate in the ROTC program leading to a military commission as an officer. With the termination of the Cadet Corp and the introduction of women to the College as regular students, these events caused a sharp decline in the number of student participating in the military classes. However many students continued to participate in the military program as well as in the drill teams of the Pershing Rifles and the Senior Platoon. These drill teams were acclaimed by many who had witnessed their performance as the best in the Nation during the prior years' performances at other locations as New York, New Orleans, and Miami. For more information on Senior Platoon, see Appendix 8.

Lynwood Banes was the president in 1956 and Jeff Reece was secretary-treasurer. Jeff was a member of the Aero Club during 1955 thru 1958. He recalled that, in addition to the two Piper Cubs, the club also had an Aeronca, N49546 and a Funk, N81190. They were still operating out of the Chicken Farm (Clemson-Central airport) as it was called at that time since the Meldeau Flying Service no longer operated the airport because of the sale of the property for the broiler operation. The club had about 25-28 members during this time and the club's flight instructor was Grady Dalton, an FAA Certified Flight Instructor of Greenwood, SC who came up on weekends to instruct those members that were still working toward solo license and/or their private pilots license. Joe Parks was the Faculty advisor for the club. During this time other pilots would use either of the aircrafts to go on cross country trips to their homes or just flying around the area enjoying the scenery. Sometime during 1954, 55 or 56, when the club was using the Chicken farm for their operations, an accident occurred. It was a rule that the pilot prior to landing, would fly down the runway to clear the chickens that sometimes would be on the runway most any time. On this particular day the chickens did not clear the runway as the student pilot was landing and a number of chickens were killed. The prop was damaged but no personnel were injured. After this incident the club moved their operation to the Anderson Airport, Anderson SC where they remained until the Cherry Bottom airstrip then under construction was completed just south of the Campus. Pete Norris was the club president during 1957-1958 and there were about 22 members. The Cherry Bottom airstrip, constructed during late 1956 or 1957 was located in the flood plain land of the Seneca River south of the College and as such it was not a good place to learn to fly as it required the pilot on final approach to clear trees and to use the slipping technique to land because of the shortness of the single runway. More experienced club members were allowed to operate from this airstrip. Most of the club's student operations however were conducted at the Anderson Airport while the Cherry Bottom airstrip was being completed by the Aero Club members and Clemson's maintenance personnel.

Carlos Lam, Jr was the president of the flying club during late 1958-1962. The faculty advisor was Don Tuck and/or Alan Cover during this time. One of the members was George G. Miler Jr. , who graduated from Clemson with a degree in Electrical Engineering in 1962. He continued his education from 1963 through 1967 as a post graduate student in Physics, Electrical Engineering, and Industrial Management. He was employed as a Research Engineer for the College of Arts and Sciences during this time. He obtained his Masters Degree in Electrical Engineering from Florida Institute of Technology in 1968. He attended a number of Bendix and King Avionics schools and other technical schools. He was employed by Harris Semiconductor on a number of research products from 1968 through 1972 and earlier he had founded his own business George Miler, Inc (GMI) in 1970 and located his business at the Greenville, S,C. airport. His company had two divisions, Greenville Avionics (FAA Repair Station 709-11) and MilerTronics. The Greenville Avionics was engaged in sales, installation and service of aircraft radio equipment. Their products were weather radar, altimeters, radio navigation equipment, automatic direction finders, transponders, auto pilots, distance measuring equipment, and communication equipment. MilerTronics was engaged in electronics and computer research and development. The emphasis was on the design and development of custom digital systems and complex data communication systems. George became a private pilot with multi-engine ratings and an instrument rating and flew in support of airport navigation systems in SC. He logged almost 4000 flying hours and was a very successful business man. His early interest and his association with the flying club played a very important part in determining his career in which he distinguished himself in such a manner with his creativeness that after his death in 1998 he was inducted into the SC Aviation Hall of Fame in 2003.

Chapter 5

The Cherry Bottom Airstrip
1956 - 1967

Sometime in late 1956 the Aero Club got permission to build an airstrip south of the college on land that was owned by Clemson College and known as the Cherry Bottom area. The Aero Club members, armed with their chainsaws proceeded, with the help of college personnel, to clear a 2000 foot runway with a 75 foot width. Some of the Aero Club members and the other people installed a fuel tank and associated pumping equipment for their use. To the north of the strip was a railroad and a road. The Aero Club had two Piper Cubs and a Funk when it begin to operated from the Cherry Bottom airstrip. One interesting incident that was reported to the author by a former club member was that the Funk, being flown either by Dr. Jud Hair (Faculty Advisor) and/or Don Clark, took off during a hot summer day and during their climb out they could not clear some power lines and the aircraft struck them. There was minor damage to the aircraft but they landed the aircraft on the nearby road. The aircraft was later flown to the Anderson airport by an instructor for repairs. The Funk later was substantially damaged in an aborted takeoff due to an engine failure.

Professor Ed Byers of the Clemson Faculty had a Fairchild 24W61A which he bought from Ben Johnson of the Anderson Airport. He kept his aircraft at the Cherry Bottom airstrip and during the summer months he would fly to a job with Lockheed Aircraft in Marettia, Ga. on Sundays and return to Cherry Bottom on Friday. He recalls about his encounters with those summer thunderstorm while flying those commuter flights. Professor Alan Cover also had a Comanche aircraft that he flew from the Cherry Bottom airstrip.

Mike Smith came to attend Graduate School at Clemson in the summer of 1961. He became a member of the flying club and was the Treasurer during 1961. At this time the Aero Club was almost in a state of extinction as there were only about 5 active members. He remained in the club till 1963. The club members decided to sell the Funk and the last of the Piper J-3 Cubs and to use the funds to purchase a used Piper Colt PA-108-11, N4943Z. Sometime in 1963 the club acquired a Taylorcraft, N95809 for their use. By the end of 1963 the club membership was 45. Mike says he has fond memories while a club member and instructor. He remembers a student that he flew with quite a lot. He taught him cross country techniques and how well the Piper Colt could be flown without power from an altitude of 3000 feet to a safe landing. Mike, after leaving Clemson, became a flight test engineer at the Raspert Flight Research Center at Mississippi State University and now has accumulated over 6000 hours. He also owns and flies a ISKRA TA-11 Polish Jet trainer.

Dr Smith is a senior aviation executive with over 42 years of experience in aircraft development and certification and has been a pioneer in the field of composite material processes development and certification methods for General Aviation (GA) and special use aircraft. He has been involved in six light plane aircraft development and certification programs and has performed extensive research and development for NASA, DOD, FAA, USDA, and the U.S. Aircraft industry. He worked for the Raspet Flight Research Center at Mississippi State University for eight years and operated a consulting engineering business for twenty years.

Dr. Smith has served as President of Global Aircraft Corporation since 1993 where he managed the development of an all composite GA airplane and developed a revolutionary composite propeller for GA airplanes that is more efficient and quieter than typical metal propellers.

Dr Smith is a member of the American Institute of Aeronautics and Astronautics, the Society of Automotive Engineers, the National Society of Professional Engineers, the Mississippi Society of Professional Engineers, and is a registered professional engineer in the State of Mississippi. He has served on numerous technical committees in AIAA and NASA programs and has served as a consultant to the U.S. and Canadian governments regarding aircraft safety and operations. Additional information about Dr. Smith can be seen in Chapter 13.

Another member of the Aero club during this time was Danielio R. Lazzilo, a student from Venezuela, who also has fond memories of his association with Clemson and the Aero Club. He joined the club as soon as he came to Clemson to work toward a degree in Animal Husbandry. His flight instructor at the Anderson Airport was a cigar smoking part time preacher and flight instructor. He was a character but a great teacher. He took his flight training in the clubs PA-108-11 Piper Colt aircraft, N4943Z. He was graduated with a Degree in Animal Production as his major and a minor in Agriculture in 1963. He returned to Venezuela and accepted a job with the Venezuela Central Union in Maracay, remaining with them for three years. He later accepted a job with CATANA, a tobacco company, an affiliate of Philip Morris, remaining with them for 20 years and later retiring as a Director when he reached 60 years of age. After a short period of retirement he accepted a job in 2001 as Professor in Agricultural Crops, Ministerio de la Defensa. He has several publications to his credit and has also published a book "Some Modern Breeds and Types of Bovine Cattle", published by the Agronomy School Library of the Venezuela Central University(VCN). He completed additional education and training at the Universidad de Caeabobo, Valencia in Adminsttracion de Fincas in 1976 and later at the Unstituto de Estudios Superiores de Administracion (IESA), at Caracas in XProgram Ejectutivo in 1981. He is also fluent in Spanish and English and has a fair working knowledge in Portuguese, Italian, and French.

His research and development activities during his Industry and government career were as follows:

Breeding and improvement of a crossbreed line of pigs adapted to the tropical conditions IAP, Agronomy School, UCV.

Swine feeding standards verification in tropical conditions; Institute of Animal Production, Agronomy School, UCV.

Prasitic effects in swine feed conversion ; Institute of Animal Production, Agronomy School, UCV.

Comparative effects of some swine parasiticides; Institute of Animal Production,/MAC

Davis University technique of Rumen fistula applied to cattle and sheep for digestibility studies; Institute of Animal Production , School of Agronomy and Veterinary Medicine, UVC.

Effects of organochlorines insecticides on N, tobacum cured leaf, reflected in chemical analysis; C.A. Tabacalera National.

Comparative studies of chemical vs mechanical methods on N, tobacum sucker control,; C.A. Tabacalera National.

Flue-cured and Burley Tobacco varities competion trial; C.A.Tabacalera, National.
New buying system to classfy cured burley and flue-cured leaf; C.A.Tabacalera National.
Project to integrate a poultry farm operation for El Tigre Farm; Corpoindustria, Maracay.

He has continued to fly, has owned a number of aircraft, some of them were a Cessna 177, Beech BE-33, Cessna 206, and a Cessna 337 and has flow over most of Venezuela, the Caribbean Islands, Costa Rica, and even to Bolivia. He founded the Maracay Aero Club in 1975 and has served four times as President. For founding the Maracay Aero Club and for conducting a number of search and rescue missions to locate downed civilian and military aircraft, or flying missions to move people from areas of distress to safe areas, and for missions for searching for lost aircraft, he was awarded in 1981 the Venezuela Air Force Cross. He stopped flying in 1987 as it became too expensive to own airplanes and fly them with the Venezuela economy the way it was in 1987. He has flown over 3000 hours in his Cessna 206 and the Cessna 337 Skymaster.

Chapter 6

A New Airport for the Clemson Area *1960 - 1967*

As early as 1958 State Senator Marshall Parker of the Oconee Delegation proposed that Federal funds be made available for the construction of an airport at Clemson. He introduced a bill in 1958 authorizing the Clemson Board of Trustees to enter into federal grant agreements under Federal Airport Act and to enter similar agreements with the State for receipt of funds available for public airport purposes. Clemson College made application in early 1958 to the Federal Aviation Agency requesting funds after they had offered a tract of land, about 155 acres, near the Fant Grove section of Anderson County about one mile from the Cherry Bottom area south of the College. Under the law at that time if the Federal Government approves the site then construction can begin. However the State and the county must put up funds with the Federal grant before the airport construction can begin. Proposals for the site survey of the Fant Grove area was released and an engineering firm from Wilmington, NC was the low bidder for conducting the survey which included the cost for grading, paving, and building of the required airport facilities. However Federal Funds were not available at this time and the construction was placed on hold. Senator Parker would not give up hope and so in 1961 the State Budget and Control Board released funds for another preliminary survey to be conducted for locating an airport that would serve both Oconee County and Clemson College. The proposed airport was to be composed of two runways of 3400 feet in length and 250 feet wide and a paved runway of 3200 feet and 75 feet wide. Senator Parker suggested a site near the Shiloh Road east of the town of Seneca. This new site was located on Clemson property in Oconee County and the flooding of the Cherry Bottom Area by the Hartwell Dam project would inundate the present Cherry Bottom airstrip being used by Clemson. Clemson agreed to give an experimental peach orchard (Kible Peach Orchard) about four miles from Seneca for an airport to serve both Oconee County and Clemson College area. About this time the Federal Aeronautics Adminstration (FAA) announced that new airports could be built and that an airport could be located in Oconee County. Funds became available from the Federal government and from the State and county.

The SC Aeronautics Commission then agreed to build an airport that would be located in Oconee County. Construction of the airport was started in 1962. The airport was finally completed in 1964, funded by Federal and State funds. The airport had a 3000 foot paved runway but no hangars,navigation aids, or lighting. It was owned by the SC Aeronautics commission and operated by a Fixed Base Operator (FBO). Reid Garrison became the first FOB around July 1965 at the new Clemson-Oconee airport as it was called . Garrison operated the airport as FBO until sometime in early 1978. There were six aircraft in the beginning, three owned by the FBO and three owned by other persons or Clubs. There were 16 tie-down spaces for airplanes along the parking apron. The runway was initially grass but was now paved with asphalt and the runway length was increased to 4800 feet paid for with State Funds later in 1966 and additional T-hangars for aircraft were planned by Garrison as demand for hangar space increased.

Reid Garrison, graduated from Clemson A & M College in 1960 with a degree in Industrial Education and was commissioned as a 2nd Lt thru the Army ROTC. He entered the U.S. Army Aviation and was active for 23 years (active and reserve) as a pilot and as a helicopter pilot. He had worked at Carolina Aero Service at the Anderson Airport earlier as a line boy and mechanic while attending college and while working there in 1957 he obtained his private pilots license and later became a Certified Flight Instructor (CFI). After he became the FBO in early 1965 at the Oconee-Clemson airport he in 1966 was awarded an ROTC Flight Instruction Contract (FIP) which provided Clemson Army ROTC and Air Force ROTC students flight instruction leading to their private license. More details on the FIP program are available in the appendices. There were about 32 students enrolled (22 AFROTC and 10 AROTC) initially. One of these AFROTC students was now retired USAF Col Alton C Whitley and more details on this individual can be found in Chapter 13. Garrison also offered a charter service using his single engine aircraft and later a multi-engine aircraft which he owned.

There were two flying clubs that used the Oconee-Clemson airport facilities, the Clemson Aero Club and the Fort Hill Flying Club. The Fort Hill Flying Club had about 30 plus members sometime after they were organized, many of them at the time were student pilots who began working on their initial solo requirements, and later their private pilot licenses. They had a Cessna 150 and a Cherokee 140 aircraft. They remained in operation for about five years. John E. Harper, a Clemson graduate and Korean War Veteran was one of the members joining around January, 1967, soloing, and then getting his private pilots license later in 1967. There were finally about 25 to 30 aircraft (FBO, Flying Clubs, and personal aircraft) that were based at the airport during the late 60's.

The Aero Club had its own Cessna 150 aircraft and they also used the FBO's instructors for a while but the arrangement didn't work out too well so the Aero Club moved their operations to the Anderson airport for a while. The club offered their members flight instruction and also ground school instruction. The President of the 1964 Aero Club was R J. Rutland and there were about 25 members. They had a Taylorcraft, N95809 that they used as their training aircraft.

Clemson A & M College is the land grant college of SC but as the technology revolution that started with the Russian launch of their Sputnik spacecraft and the beginning of the Space Age there was a considerable change taking place at Clemson. Many changes were made to the faculty and many old buildings were renovated and many new buildings were added. The total enrollment was on the rise and with Clemson's involvement now in more research, training programs, and other services offered to the State and to the students of Clemson A & M College, the land grant college became Clemson University on 11 July 1963. Since Clemson had been a land grant college and an all male college with the mandatory requirement that all students must participate in military training, the number of ROTC students began to declined from 1955 to the early 60's however there were some students that were still interested in military careers. The ROTC program was not mandatory now but was still offered to those students and by now more and more women were being admitted to the school as students. It was during the late 60 and the early 70's that women became members of the Aero Club.

Herbert J. Morgan, Clemson Class 51 and a State representative from Oconee County also recognized the potential for a modern airport to attract business to Oconee County and he introduced a resolution that would turn over the operation of the Clemson-Oconee airport to Oconee County. The resolution introduced in 1967 was passed and Oconee County assumed ownership of the State owned airport. Morgan created an advisory board and appointed John E. Harper, Dr. Jake Booker, Ross Davis and Tom Campbell to the first board. Later three more members, Jerry Wigington, Michael Williman, and Terry Wiliams were added to the advisory board.

Jerry Wigington, a local businessman from Walhalla, SC was one of the general aviation aircraft owners. Jerry had a Piper J-3 Cub, a Cessna 140, a Cessna 180, and a Bucker Jungman (a German acrobatic airplane) at one time. There were about 10 to 30 aircraft based at the airport at any one time during the 60's not including the Aero Club's aircraft and the Fort Hill Flying Club.

The president of the Aero Club during 1967 thru 1968 was Steve Coons. When he joined the club the club had the Taylorcraft they had purchased earlier and he flew it and took instruction from an instructor named David. Steve obtained his private pilot license in the Taylorcraft. The club later sold the Taylorcraft and purchased a Cessna 150, N86686, during his term as the President. He was graduated in 1969 with a degree in Mechanical Engineering. He stopped flying in the 70's as family priorities came first. He however remembers the time when he and his instructor were flying in the Taylorcraft and a small fuel leak developed from the fuel connection on the bottom of the fuel tank that was located in the cabin area and the line that passed into the firewall. The instructor just cracked the cabin door to let the fumes dissipate and we continued the flight. The aircraft did not want to fly straight with the door open. The rest of the flight was no problems.

Clemson student Gaines Liner joined the club in the late 60's while attending Clemson as a Graduate Student. He soloed the Club's Cessna 150 and after getting his Ph.D he later got his private pilot license at Charlotte's Douglas Airport. He later became an Associate Professor of Economics at UNC Charlotte. Learning to fly was, as he said, was one of the highlights in my career.

For a while another student was Milt I. Turner who was in the Aero club during the period of 1963-1964. He took his flight training at the Carolina Aero Service facilities located at the Anderson Airport under the Army ROTC program and received his Private Pilots license. This program required that the participants after completion and graduation that they would be required to serve for at least three years in the Army if they wanted to continue their Army flight training and receive their Army Pilots Wings. Before graduating from Clemson in 1964 he joined the Aero Club and flew the Club's Taylorcraft from the Cherry Bottom airstrip which was then in operation. After graduation, as required, he entered service and served two years but did not go to go the Army flight school. He remembered that during one flight in the Club's aircraft he took a friend flying with him. After flying around he then decided to demonstrate a stall to his friend. Once into stall attitude a wing dropped and the aircraft fell into a spin but he was able to get it out and the friend thought that was fun. After he returned from service he continued his flight training under the GI Bill while living in Shreveport, LA. He received his commercial and instrument ratings under this program. He later was a member of a flying club in Nashville, TN for about seven years and used the club's aircraft for business as well as pleasure. He later became a Sales Representative for Universal Solutions, Business Partners with Avaya, a spin off of the telephone business from Lucent Technologies. He operates his own telecom business now in Charlestown, SC.

During the 60's a number of new buildings and the renovation of former buildings on the campus of Clemson A & M College was on the increase. The Freeman Building was slated for renovation and so the "little 372" that was hanging from the rafter in the woodshop was slated to be scrapped if a final resting place could not be found. When the remaining seven original members that built "little 372" heard of the little aircraft's plight they rushed to her aid. Each of them contributed $100.00 to help in the restoration of the little aircraft and to find a final hangar for it. The little aircraft was removed from its resting place and it was transported to the Anderson Airport where it was completely refurbished by Ben Johnson in its silver gray and red trim as when it was first built. It was then returned to the Clemson area and remained there until Doff Overton III, owner of the Wings and Wheels Museum of Santee, SC, agreed to place it in with his collection of old and valuable aircraft. It remained with the museum when the museum was relocated to Lakeland, Florida. Overton decided to close the museum and to auction off all the aircraft and other items. The return of 372 to Clemson was through the efforts of Dr. Ed Byers and Heyward Douglas. Little "372" remained at Clemson for several years in Dr. Byar's hangar untill she was loaned to the SC Museum in Columbia, SC where she now hangs from the ceiling as if in flight.

During the late 50's and early 60's Professor Joe Rostron of Clemson's Civil Engineering Department was active in his flying career as an advisor to the flying Club but he also used a Fairchild F-24 aircraft, owned by Dr. Ed Byars, which they had equipped with a surplus K-17 (9" X 9") large format camera for aerial photo work for Clemson College. The camera was used to take photos for Clemson forestry management and photogrammetry classes as well as for other engineering uses. The camera was a good photo device but without a view finder it was difficult to fly a straight line with the radial engine blocking the forward view. However in the spring of 1960 Dr. Byers took a job as Chairman of Mechanical Engineering and Mechanics Department at the West Virginia University and took his aircraft with him. Prof. Rostron continued to use the Fairchild camera until he finished the job of photographing the proposed Hartwell Lake area before the construction of the project was started. Ben Johnson of Carolina Aero Service of Anderson, SC had a Cessna 172 (N5812A) with a tricycle gear that made it easy for it to be used as a aerial photography ship. Joe and Bill Shane (Forestry Department Photogrammery Prof.) designed a fiber glass camera pod that could be mounted on the belly of the aircraft. Joe flew the leased aircraft and used this pod to take aerial photos. For the next 25 years or so (well past Joe's retirement) they continued to take aerial photos for the College as well as numerous SC State Agencies as requested. Bill Shane later purchased the aircraft around 1990 for aerial photo usage but it was later destroyed in a crash by another individual while flying a photo mission thus ending Clemson's aerial photo program.

Nick McNair entered Clemson as a student in September 1966 and joined the flying club in 1967. He was a member until 1969 and took his training at the Clemson-Oconee Airport. He remembers that Reid Garrison was the FBO at that time. Nick soloed the club's Cessna 150 and later got his private pilot license. He remembers that it was pretty cheap to fly as the cost for the rental of the C-150 was $8.00 per hour and the instructors fee was $4.00 per hour if he paid the instructor. At that time there was a constant battle with the FBO and the flying club because the FBO wanted the Club to use his instructors for their flight training. For this reason the club moved their operations to the Anderson Airport south of Clemson. After graduating he received his commission on May 8, 1970 as a 2nd Lt Army Medical Service Corps just prior to graduation. He entered military service and in April 1971 he entered Army Helicopter Flight School, completed it, and became a Helicopter pilot. He flew actively in the Army for about 16 years of his 21 years in the Army. He is in a partnership with other people and they have a Piper Lance , PA-23 RT 300, and he still flies as much as he has time for. The aero club was the begining of a life long enjoyment of flying for him. He said that he really enjoyed flying in the Clemson area and enjoying the scenery. One incident, not related to flying but interesting, was that he was in a hurry to get to a flying lesson on time after his Physic class and so he broke in the food line at the Harcomb Commons Dining Hall in front of a young lady. When he gave her the reason she commented that her dad owned an aircraft. Later he asked her for a date and they were married after graduating in 1970. They are still happily married for over 33 years.

Carlos Lam Jr was the president of the flying club from 1967 thru 1968 while he was attending graduate school at Clemson. Some of the members of the club were B. Sency. A. Buckalew, N. McNair, F. Johnson, B. Henderson, C. Freeman, and M. Rogers. During 1969 Guy Cannavagio was the president and in 1970 Carlos again was the president.

Chapter 7

A NAME CHANGE FOR THE CLEMSON AERO CLUB *1967 - 1973*

During 1971-1973 the Aero Club's President was Don Fraser and he was later succeeded by Ross Sanders. The club's faculty advisor during 1971-1973 was Professor J.B. Todd who himself owned a Cessna 172, N2650L. The Club had about 20 members and they owned two aircraft at that time, a Cessna 150, N8668G and an Aeronca 7AC, N4455E. The club members rented the Aeronca to themselves at $5.00 per hour including the fuel. The Club origionally purchased the Aeronca on October 2, 1971 from "The Red Barrons" a group of professors who had bought it themselves from Ruth L. Lam, wife of Carlos Lam, Jr who had been the president of the flying club earlier during 1958 thru 1962. The Aeronca was flown to Spartanburg in the spring of 1972 for its annual maintenance check and a cylinder was found to be somewhat low during the normal compression checks but the aircraft was certified for flight. Don Fraser was the pilot for the return trip back to Clemson, however after a period of time in the flight and after he had passed Norris, SC the engine made a "large pop" and the aircraft began to descend as the engine began to loose power and could not maintain altitude. There was really no place to land at that time since he was now between the Clemson airport and the Pickens Airport. Below was a large parking lot of a manufacturing plant but since he still had quite a bit of altitude he pressed on toward Clemson. He landed safely at the Clemson airport and as he landed the prop stopped. After inspection of the engine it was noted that a spark plug had not been properly torqued and that is why it blew out. Since the club did not have enough money to be able to repair the engine and the members decided to sell the aircraft. It was sold in

September 1972.

This same Aeronca 7AC N4455E after being owned by a number of other owners over the years was later bought by Don Fraiser and is still in excellent condition being flown almost daily by Don from the private airstrip (Eagle Ridge Airport SC24) located three miles east of Seneca, SC) where he and a number of other pilots have built homes there at the airport.

During the period of 1979-1984 The club operated from the Oconee County Airport and had the following aircraft: a Cessna 150, a Cessna 152, and a Cessna 172. The faculty advisor for the club was Professor B.J. Todd. The club lost its non-profit tax exempt status sometime around early 1980 due to a misunderstanding on some correspondence exchanges between the club officers and the IRS. This problem resulted in the grounding of the club's aircraft. A solution was achieved to the satisfaction of all parties and the tax exempt status was reinstated but it resulted in a requirement to change the name of the club. The club new approved name was Aeronautical Association. This required name change also helped the club. Prior to the IRS problem "Clubs" at Clemson were not eligible to receive funds to help them in their operation costs but the name change made them now eligible. Sometime during the 70's the club moved their operations to the Anderson County Airport for a short period of time. They later returned to the Oconee County airport while the FBO was a company called Tiger Aviation, managed by Alan Laudermilk. Laudermilk was the FBO during 1975 through 1981. The club had 10 active members during 1979.

One member, Greg McDowell, was a married Vietnam veteran. While attending Clemson he was a graduate student assistant to Dr. J. Page Crouch, head of the Graphic Arts Department for Industrial Engineering. His wife saw the flying club's Cessna 172 aircraft that had landed on the highway near the Clemson Library and had been taxied by an instructor to an area on Bowman Field in front of Tillman Hall. It was a custom of the flying club during the Freshman Orientation Day to show the students their airplanes and hope that some of the students would be interested in joining the club and become pilots. He decided to give her husband a present of flying lessons. He remembers that the cost for a flying lesson was $23.00(Fuel-$16.00 and the Instructor--$7.00 per hour). After he had completed about six hours of dual in the club's Cessna 150 aircraft he was ready for his solo flight. He had been having trouble before his solo flight with his landings--getting the correct attitude during the landing phase. The instructor told him that we were not making carrier landings as we do not have a tail hook. Some of his friends gave him a nickname "Tailhook". After his seventh hour of dual solo he was making his third landing, having completed two others successfully, and was cut out of the approach pattern by a twin engine aircraft that had flown to the airport from Anderson. This caused him to abort and come around for another landing. He then landed ok. After he landed he lost his shirt tail as was customary when you solo. His instructor cut out the back of his shirt and then his instructor added the date, who, when, and where and then autographed it with" Nice Go Around Mac". He saved that shirt tail and it is now hanging on the chalkboard in the Senior High School classroom where he teaches at Hudson Falls, NY.

At the same time he was learning to fly there was another student, Susan Pagent, who later got her private pilot's certificate. Susan was the second femail club member. Mac did not get to finish his training for his private certificate because of family and other responsibilities while in Graduate School. He now has instilled in his son the love of aviation and builds electric powered aircraft with his son. Being able to look back and remember when he soloed it has been a blessing that he will never forget and is grateful for the flying club for his experiences.

Another member of the flying club was Kevin Prufer who enrolled in Clemson in 1977 to study Economics. He joined the club in the spring of 1978. He already had his private certificate, obtained while taking flying instruction from Hillman Flying Service in

Roanoke, Va. He was very much interested in flying for an airline. During 1977 he worked for Reid Garrison at the Anderson Airport. Between his junior and senior years he began working toward his instrument, commercial, and multi-engine ratings at the Colgan Flying Service in Manassas, Va. and received his certified flight instructor rating. He receiver the instrument, multi-engine ratings and completed most of the requirements for his CFI. He returned to Clemson in the fall of 1978 where Heyward Douglas endorsed him for his CFI check ride. He flew to Columbia, SC and there completed his requirements. He then began instructing for the Clemson Flying Club and for Alan Lowdermilk, the FBO at Tiger National Airways at the Clemson Oconee airport. After he was graduated from Clemson in May 1979 with a degree in Economics and a minor in Accounting he moved to Charlesville, Va. There he was hired as a primary flight instructor and Air Ambulance pilot at Cardinal Aviation there in Charlesville, Va. flying Piper Navajos and Swearinger Metroliners. He flew with Cardinal for a while and in 1981 was hired by Piedmont Airlines as a Co-Pilot. Piedmont was merged with USAir in 1988 and was later renamed US Airways. In 1985 he was promoted to Captain. During his career he has flown the B737-200, B737-300, B737-400, B727, and the Airbus 319, 320, and the 321. He has been a Captain for 21 years and has also worked as an instructor on the B737-200. Being a member of the Airline Pilots Association he has been a First Officer Representative, MEC Vice Chairman, and MEC Secretary and Treasurer.

He relates that one of his students was an older woman there in the Clemson area who at one time was a pioneer pilot and she wanted to reacquaint herself with aviation after a long absence. She wanted to solo and to learn about radio communication and navigation. He said really he enjoyed working with her.

Chapter 8

A gift to the Flying Club *1980 - 1990*

During this time frame the club was a fairly small group of people and were close to extinction again. They managed to gain in membership in the early 80's. Chip Houck was the President of the club in 1981. They had a Cessna 150 and later they acquired a Cessna 172, N737RP sometime around 1983. They were also given a donation in late 1982 or early 1983 of a Grumman AA-5 Tiger by retired USAF Col. Dallas Sherman, a former CU Alumni and a charter member of the 1927 Clemson Aero Club. Col Sherman worked with Dr. Ed Byars to get the aircraft ownership transferred to the club. The Tiger was flown by a number of members until sometime in 1985 when it was involved in a night-time landing accident. It was not repaired at that time and was later sold. The aircraft was later repaired and has had a number of owners, and is still flying today. The President during 1985 was David Bell and Vice President was Scott Seible. Later the President was Kevin Redmon for the year 1985-1986. Kevin remembers the nickname "Crosswind Kevin" because of the many touch and go landings that his primary instructor Hans Scott had him do at the Anderson Airport where the wind seemed to always be blowing at 90 degrees to the active runway. During the period from 1985 to 1986 there were about 30 to

35 club members and Joe Boykin was the Faculty Advisor. The FBO during 1981 through 1990 was James Thomason III and his wife.

There were at least four Globe Swift aircraft based at the airport during this time as well as a number of other private aircraft that were based there also.

One of the members of the club, Andy Applegate, obtained his private pilots license in April 1981 and continued to fly with the club till he was graduated in 1985. While at Clemson he was an AFROTC member and after graduation he entered the USAF as a 2nd Lt and went through pilot training as a student officer, remaining at Laughlin AFB, Del Rio, Texas as a T-38 instructor pilot for 15 years. He later transitioned to the C-130 and was stationed at Yakota, Japan. He returned to the states and was assigned as a C-130 exchange pilot with the Canadian Air Force at CFB, Trenton, Ontario. His last duty station was at Little Rock AFB, Arkansas. After release from active duty in 1998 he began flying for Northwest Airlines as a First Officer. Around 2003 the airlines were cutting back on their services and began to furlough pilots. Andy was furloughed since he was a new hire. Andy then remained in the USAF Reserve acting as a Liaison Officer to the Arkansas Civil Air Patrol and is now working for a flight/operations consulting company in Little Rock, Arkansas. He is scheduled to be recalled by Northwest in early 2004. As of this date Andy is now back flying and the flying club has had quite an influence on Andy's and his career.

Another Clemson Graduate student, Dr. Dan Holden, gives us a view of his experiences at the Clemson Oconee Airport as it was called during his stay at Clemson while working toward a PhD in Physics.

"My first logbook entry that contains CEU is dated August 25th, 1984. I was to begin my first semester as a Ph.D. student in physics. It was my second attempt, however, to fly into Clemson airport. Earlier in the spring I flew from my home field in Roswell, NM, yeah, THAT Roswell, to check out the Clemson campus. Weather conspired to give me, a VFR pilot, a little IMC, a diversion to Nashville and a drizzly drive in a rental car. When I finally did arrive about ten miles out, looking for the airport and talking on the Unicom to the FBO, I was a bit unsettled by the "good" visibility of ten miles. I had all of 144 hours, most of which had been logged in the Southwest, where visibilities of 100 miles or more were the norm. This was going to take some getting used to! Subsequent puddle jumping and practicing in the pattern at near sea level also gave me some surprises as all of the 235 horses were available to me in my Cherokee. I had bought the plane in New Mexico before I ever took a lesson. I was that confident that I would get my certificate and love flying. 8514W is now gone, but my love of flying is as strong as ever.

During that first rollout, I asked the FBO where I should park. He asked back, "How long you planning on staying?" I said, "Well, I hope to get done in three years, what do you think?" He laughed and directed me over to a spot that I would call my own. I flew out of Clemson for about two years until my meager graduate student stipend caught up with the reality of owning a high performance aircraft. I had to sell '14W, which closed the door on flying. One of my fondest memories at Clemson led to opening that aviation door again some ten years later. There were several local owned Globe Swifts at the airport field. I knew nothing about them except that they looked terrific and on many Saturdays a few of them would form up and blast off to breakfast somewhere. I met one of the "Swifters" as they called themselves, through a minor incident. I was moving my Cherokee over to the wash rack, and passed the wing over the wing of a Swift. My wing tank drain valve just caught the top of the wing tip on one of the Swifts, leaving a long gouge in the paint. Naturally, it was the one with the nicest paint. True to the style of the south, the mechanic walked over and said, "Well, it looks like you're going to learn how to make a paint repair." He didn't call the owner right away, though a day or so later, the owner came by while I was sanding and masking and asked me what I was doing to his plane. I explained everything to him, conscious of my mostly empty pockets, hoping he wasn't going to make me pay too much. He simply looked at my work for a minute and made a comment about checking it out later. The next Saturday after I had finished, he came by and said, "Hey, you ever been up in a Swift?" I said no, or something to that effect and he offered that they were going to breakfast. I could ride with him and he'd even buy! On the way over he asked me, "Hey you wanna do a roll?" I said, "Sure! How do you do it?", while starting to reach for the yoke. He looked over with a glance that told me very clearly that I would just be along for the ride on this maneuver. We (he) did a few glorious rolls on the way to breakfast and I was hooked: both on Swifts and on aerobatics. Though I don't remember his name, he gave me that gift of a generous nature that most pilots have, the willingness to share their passion for aviation.

I bought my own Swift in '98, in which I still share many rides with aviators and non-aviators, alike. And to satisfy an aerobatic urge that the Swift could not quite satisfy, I later put together a basket case Pitts aircraft, rebuilt it, and still pursue aerobatics for fun and competition.

My other fond memory of flying at Clemson was during the first Homecoming game during football season. I was asked if I wanted to come help park planes on a Saturday. There was to be a free lunch involved, so I naturally said I would. Not even knowing what the ACC was, I had no idea what was to follow. I assumed a few Cessna and Piper aircraft would show up, maybe the odd Mooney. I was completely blown away by not only by the sheer volume, (there were well over 100 planes) but also by the aircraft types and that many of them were painted in Clemson colors. There were Twin Commanders, Senecas, pressurized twins, and even a bizjet with a huge tiger paw painted on the tail. The small pilot lounge filled with professional pilots, who watched the game on TV, while their passengers saw it live at Death Valley. I came to a new realization of what IPTAY was all about!"

During the 80's the AFROTC cadets that wanted to be USAF pilots were offered a Flight Screening Program (FSP) that was conducted at Columbus AFB, MS. The program was used to screen out those cadets that did not really have the aptitude to be a USAF pilot and would in most cases not be able to complete the regular USAF pilot training course. AFROTC cadets that completed the FSP course successfully and later entered active service, completed the Undergraduate Pilot Training (UPT) progrram successfully, and won their USAF silver wings. Some of the Clemson AFROTC cadets were as follows:
Mike Frazier, Randy Abbott, Dan Ball, Steve Barbour, Brian Ratchford, Brad McCoy, Scott Bowen, and Kevin Earnest.

Another club member, Dr. Richard Figliola, joined the club in 1987 and soloed in September 1987 in Cessna C-152, N65407, got his private license, an Instructor license, and later his instrument ticket. He flew a number of students in 1999 to the Experimental Aircraft Association (EAA) Convention held in Oshkosh, WI. He then in 1996 purchased a Cessna 182 for his personal use. He has flown over most of the lower southeastern states and Canada. He has also taught an aircraft design course for a number of years and he was Chairman of Department of Mechanical Engineering from 1995-2002. He remained active in the club till 1996 and still attends club meetings occasionally although not a member.

One thing Dr. Figliola remembers is how the airport was operated during the time around 1987 until 1997.The FBO for the airport was Ray Smith of Uplands Aviation. The airport had a dingy block and metal building back then with an attached maintenance hangar and classroom. There was a trailer home behind the main building. The lobby was decorated with dozens of "solo" shirts and an big Aeronautical chart on one wall. The airport was a great place to hang out, particularly on Friday afternoons. Many would sit on the front porch in the rocking chairs and watch student pilots do their crash and dashes and the sitters would do hangar flying. Often someone would try their hand at acrobatics. Another event at the airport was a bomb drop contest and a balloon burst contest. The aircraft would fly over the airport and they would try to see how close that they could come to a target located on the runway by dropping a sack of flour from the aircraft. Also they would release a helium filled balloon and the aircraft would fly down the runway at about 1000 feet and try to burst the balloon. During these events an Instructor would be in the aircraft for safety reasons. About 6:00 pm we would then head out for pizza and beer. Ray was pretty layed back in his operation and there was always an airplane in the hangar being resurrected. Ray restored a Taylorcraft, NC8543 and I got a few hours in it. Dr. John Collier, an Agricultural Engineering Professor, taught many of the ground schools in the 1980's and early 1990's. He also restored a Taylorcraft L-2 and now restores other aircraft at the Eagle Ridge private residential airstrip just north west of the present Oconee Clemson Airport. Retired Dr. Ed Byars and the other residents and aircraft owners who live there restore old aircraft.

Another member, John D. Leaphart, a commercial Jet pilot and instructor, got his private, commercial, and instrument rating through Clemson Aeronautical Association as it was called at that time. He stated that the training at Clemson was second to none in that the instructor pool was diverse and the aircraft that were maintained in excellent condition. It was an excellent facility for training since the students could make multiple touch and goes without too much traffic. The fact that the Anderson Airport was close by with it's FAA Flight Service Facility was an asset for student training. In addition the FAA Greer Approach Control was always willing to provide the services of a Class C airport which was very useful in the more complicated phase of our training program.

Now as an instructor in the Citation Jets, he stated that "I have seen quite a few pilots who went through the "professional" pilot programs at Emory-Riddle, Flight Safety, and the University of North Dakota and I feel that the flight training and ground school training we got at Clemson was as good or better than anything that these schools crank outÉ.and we had football and tailgating. There is no doubt that the training that I got through the Clemson Aeronautical Association is the foundation for the success that I enjoy in my flying career."

Another student was Louis Jordan and his instructor was Heyward Douglas and then Tony Barr. He remembers the great times he had while working toward and receiving his Private Certificate and then doing much flying around the lake at Clemson. He recalls that the club had a Cessna 150 and the Grumman Tiger that was given to the club. The cost of flying for the Cessna 150 was $15.00 per hour and $22.00 per hour for the Tiger. Only private pilots were allowed to fly the Tiger. Later during his aviation career he owned a Mooney MG-20.

Another student was Cole Bristow who attended Clemson from 1986 through 1990 when he graduated. Although he was already a student pilot he became a member of the flying club his Senior year and took more flying instructions from Gary Utz, another student who was a certified Flight Instructor (CFI). He completed the training for his Private Certificate before he graduated. He also remembers the Cessna C-172 that the club had as well as the Cessna C-150. Flying it cost $25.00 per hour. A member could check out either plane for a weekend cross country and there was no minimum flight time requirement to use either aircraft. About a year after he graduated he was accepted in the Naval Flight Training program and reported to the Naval Air Station at Pensacola, Fl. for Aviation Officer's Candidate School. After receiving his commission in April of '92 he then reported to Milton, Fl. to begin his primary training to become a Naval Aviator. He completed this phase and then reported to Corpus Cristi, Texas for his advanced flight training. He was awarded his Naval Aviator's "Wings of Gold" in November of 1993. He then reported to NAS Jacksonville, Fl. for his fleet training flying the Lockheed P-3

Orion. After completion of this phase of training he the was assigned to NAS Brunswick, Maine for his first fleet assignment. He flew operationally with Patrol Squadron 8 and 11 in the Mediterranean, South America, Africa, and the Balkans during the Bosnian conflict. He requested a discharge from the Navy in 2001. He has not done any flying since, partly because it is hard to find a flying club that rents aircraft cheap as the Clemson Flying club that started him on his flying career. He said "It was a great experience to make that first solo flight and even better to get my "Wings of Gold" and the other experiences that I was able to have during my flying career".

Jeff Treece came to Clemson in 1989. He already had his Private Certificate and after hearing that there was a flying club in operation he decided to become a member. He was checked out in the Club's 172 by instructor John Boobas. John was a former Clemson football player whose knee got injured so he got his flight instructor certificate and instructed the rest of his years at Clemson. He started Jeff working on his instrument ticket and during his training Jeff got to serve at one time as a co-pilot in a King Air, (a turbo prop powered airplane). Jeff was not sure of his career path so he found out his Management professor (Terry Leap) was also a flight instructor so Terry helped Jeff finish his instrument training. Jeff recalls that he was Terry's first instrument student. The instrument check ride had an interesting twist. He had scheduled the check ride at the Greenville, SC airport in the early morning. When he got to the Clemson airport the aircraft had a coating of frost on the wings. Rather that flying the aircraft he elected to wait for at least an hour so the sun could burn off the frost. This problem meant that if he waited he would miss his appointed time for his scheduled check ride so he decided to drive to Greenville since it was only about 30 miles. He contacted Terry before he left and asked Terry if he would fly the aircraft to the Greenville airport later that morning so when he had finished his oral part of the testing he would have the club's airplane to complete the testing. The instrument instructor was impressed with his judgment and the fact that he had not attempted to fly the frosted aircraft.

Jeff mentioned that at times he would rent the club's Cessna C-172 aircraft at the county airport at Pickens, SC. to fly to his home in Pittsburgh, Pa., about 3.5 hours flight time. The plane in this case was scheduled for his flight during the Fall Break weekend but the club's aircraft was not flyable. It was in maintenance for its annual. He then called the county airport at Pickens, SC, was checked out in their C-172 by an instructor by the name of Bo Newman. "Well you know how small the aviation community is, ten years later, while flying for Air Tran Airways during my DC-9 training, Bo was my training partner" said Jeff.

Jeff, after graduating from Clemson, went to Pittsburgh and got his ratings the Multi-Engine, Commercial Flight Instructor (CFI), and his Commercial Flight Instrument Instructor (CFII). He was then hired at Ainet Express (US Check) flying for Banks(cancelled checks) throughout the Northwest. He flew Beach Barons, Navajos, and then an Aerostar. Jeff flew for about 1.5 years and then he was hired by a Federal Reserve contractor flying single pilot night freight in a Mitsubishi MU-2. That was by far his favorite "hotrod" airplane to fly. He remained there for 10 months and the moved on to Air Tran Airways.

Coming to Air Tran was a gamble at first because it was still considered a startup and they flew the old DC-9s. He started out in the -9 and then moved over to co-pilot on the brand new Boeing 717(120 passengers). He was a co-pilot for only three years and then upgraded to Captain on the Boeing 717.

"It is strange how the airline industry operates - Air Tran had been hiring a number of furloughed major airline pilots with twice Jeff's experience and strangely some of those pilots at one time became his co-pilot. Being just 34 years old one of Jeff's co-pilots was the Air Force One pilot for Bill Clinton and George Bush recently", Jeff said.

Tim Lewis attended Clemson working toward degrees in food science and later in

Mechanical Engineering. He was awarded his BS Food Science degree in 1989 and his BS in Mechanical Engineering in 1991. While at Clemson he was a member of the flying club during that time and was the president for one year. George Roland was his instructor for his Private and Instrument training. He remembers the Organization Day sponsored by the University in which the flying club was allowed to bring their aircraft to the campus. The Cessna C-152 aircraft was flown by Heyward Douglas, the Faculty Advisor and Chief Instructor for the club, He would land the aircraft on Highway 93 and then it would be taxied up the road to a position behind Brackett Hall for static display. He also remembers the airport contests that were held as mentioned earlier and the cookouts that were held after the contests. He is still flying today and has earned a commercial rating in February 2004 in addition to his other ratings. He has flown Cessna singles, a Beech Sierra, and Piper singles. Tim lives and works in Spokane, WA and really enjoys the beautiful scenery flying in the Northwest. Since they have the Cascades to the west and the Rockies to the east most of the flying is done at high altitude since the MEAs around there are 10,000 and 12,000 feet.

Scott Bowen entered Clemson and was graduated in 1988 with a BS degree in Computer Science and was active in the AFROTC program, a Distinguished Military Student, receiving his initial USAF commission as a 2nd Lt upon graduation. He then entered the USAF, completed his flight training, and was awarded his wings as a USAF pilot. He was not a member of the Clemson Flying Club while at Clemson but he made many flights with a friend, Kevin Earnest, who was a member of the flying club. Kevin later became a Captain, flying C-130 Presidential Support Aircraft during his military career. Scott was also a Thunderbird pilot flying the slot position.

The early flight experiences played a part in Scott's desire to become a pilot. More details about Scott are in in Chapter 13.

Chapter 9

Clemson Flying Club and the Oconee Regional Airport *1990 - 2000*

During the early 90's the Club's Cessna 172, N737RP sustained major damage when it was struck by a moving aircraft. The aircraft was being flown back to Clemson from Florida and while the stop was at Aiken, SC after landing and taxing into their assigned parking space, the accident occurred. Seems that a pilot attempted to hand prop his Skybolt aircraft and as it started it got away before it could be stopped and the C-172 was struck near the horizontal stabilizer and the prop did considerable damage. However it was repaired very successfully and flew many more hours for the club.

In 1990, an executive from Bendix and a former Aero Club member donated two full radio stacks for each of the club's two aircraft and has continued to help the club.

During the late 90's the club changed its name to the Clemson Flying Club. The fixed base operator (FOB) from around 1987 through 1996 was Ray Smith's operation called Uplands Aviation. Ray offered flying lessons, aircraft hangar facilities, and aircraft maintenance to the general public.

Marion Lyles, a retired Delta employee succeeded Ray Smith as FBO while employed by the county sometime around 1996 and he was a driving force in obtaining federal grant funds to build a new terminal and to get new hangar facilities. He left the airport position about 2001 to become a member of the Oconee County Council. The airport operation during Marion's term was a county controlled and not a commercial airport.

Harrison Orr was elected Oconee County Supervisor in January 1997 and began an effort to make more improvements to the airport. Wayne Rholetta from Walhalla, SC was one of the pilots and other Oconee County residents that wanted to expedite the improvement to the airport. Rholetta flew Orr to Columbia a number of times. Since the airport was still a small mostly non commercial country airport it was time to bring the airport up to modern standards. Orr made a number of trips to present his opinions for improving the airport for more commercial usage with not much success. Oconee county did not have a Comprehensive Land Use Program so that progress was not forthcoming. The airport improvement project began in June 1997 after a Comprehensive Land Use Program was developed and accepted by the Council over some local opposition but the improvement program was finally initiated and the improvements to the airport were completed in September 1998. The new terminal building included an attractive lobby, office, adminstration area, and conference room with a kitchen. The addition of a parallel taxiway was created to make the airport operations safer since there was a rise in the center of the runway and it prevented pilots of incoming aircraft from seeing taxing aircraft on to the runway. The addition of this parallel taxiway helped eliminate the potential safety problem. Three long Tee hangar buildings were also constructed west of the terminal and these buildings housed aircraft. An 8000 square foot maintenance hangar which is attached to the new terminal was also constructed on the east end. Oconee County received a significant financial assistance from the Federal Government, the State, and the local corporate community.

The February issue of the 1998 Airport/Facilities Directory showed some of the important information for the Clemson-Oconee Airport as follows:

The airport is attended Mon-Sat 1300-23Z, Sun 1400-2200Z. Parachunte jumping. Other aircraft may be operating from a private airport 1 nautical mile W-NW of the Clemson-Oconee(CEU) airport. CEU had medium intensity runway lights(MIRL), Runway End Identifier Lights(REIL), and Visual Approach Slope Indicator System(VAIS). The MIRL system of lights could be turned on by the pilot activating the Common Traffic Advisory System by clicking his microphone button. Fuel 100LL and Jet A are available. There is one runway 07 25.

Some of the Clemson Flying Club members during the 1990s were: Terry Leap, Janis Miller, Phil Quillin, Eric Lowery, Clay Schile, Grant Kipp, John Pasek, and Chuck Corrigan. Vonda Bradt, P. Ellen Bethea, Peter Donald, Lois Nice, Jeffery Marshall, Joel Robinson, Parker Willey, Josh Rogers, Rea Hodges, Bryan Rodwell, Randy Collins, Scott Ellis, Jeff Treece, Kim Cobb Gramling, W. Barlage, J. D. Leaphart, Vickie King, Taylor Sutherland, Thomas Philips, David Norrill, Johnie James, Eric Lowery, Philip King, Gary Utz, Andy Watson, Douglas Baker, Larry Leffler, George Todd, Tim Lewis, Heyward Douglass, Cole Brisko, and William York.

These are but some of the few people that the author has have been able to identify as former members of the Flying Club during the 90's.

Chapter 10

Clemson Flying Club
2000 - 2004

During 2000-2002 the club had 43 members that were active in the club. Peter Donald was President. Will York was Treasurer, and the faculty advisor was Heyward Douglas. Flight Instructors were Terry Leap and Randy Collins. The club's aircraft were a Cessna C-150 and a Cessna C-172.

During 2002 the club had 30 members that were active in the club. Will York became the President, James Mack was the Treasurer and Heyward Douglas was the Faculty Advisor. The Flight instructors were Terry Leap and Randy Collins. The Club's aircraft were a Cessna C-152, N65407and a Cessna C-172, N7377RP.

During 2003 the club had 43 members that were active in the club. William York remained as the President, James Mack, Treasurer. Heyward Douglas was the Faculty Advisor. Faculty Flight Instructors were Randy Collins and Terry Leap. Part-Time Instructors were Pat Kohout, Jason Williams, and Shane Valentin, The club's aircraft remained the same.

Dan Schmidt grew up in Charleston, SC and came to Clemson to get his degree in Computer Sciences. He was awarded his degree, and then went to work for the college in the Computer Center. Although nobody in his family had been a pilot he decided that he would join the flying club and learn to fly. After joining the club he completed his period of instruction, flying a Cessna C-150 under Pat Kohout. After about three months he received his private pilot license in December 2002. After flying the club's aircraft until March 2003 he decided to buy a 1946 Luscombe 8A, a tail dragger aircraft. This was a new and exciting flying experience for him and after about 15 more hours' instruction by Mike Walling in the taildragger he was allowed to solo it. His Instructor told him the best way to learn to fly the aildragger was to fly it as much possible. So every morning before going to work he would fly the aircraft for a least 30 minutes. By doing this he logged about 650 hours as pilot in command. He then decided that he needed a better aircraft so he purchased a 1980 Cessna 172RG, N20CU in which he got his commercial license and instrument ratings. He is now working toward his Instructors license and plans on instructing for the club.

He is active in the Flying Club but still flies the tail dragger as much as he can. He says it is the greatest. One of the big thrills he got recently in 2006 was to fly a DC-3 from the left seat or pilots seat during a takeoff, a flyby, and a landing as pilot in command supervised by an instructor who was the owner of the aircraft.

During 2004 the club had 27 members. James Mack was the President, Zack Hoffman, Vice President, Joe Braun , Treasurer, Erin Shull, Secretary, Mike Svoboda, Safety Officer, and Heyward Douglas was the Faculty Advisor. They had two Faculty Flight Instructors, Randy Collins and Terry Leap and four part time Flight Instructors, Pat Kohout, Jason Williams, Pat Hartigan, and Shane Valentine. The Club's aircraft were a Cessna C-152 , N65407 and a Cessna C-172, N737RP.

The Cessna C-172, N737RP was damaged by a large thunderstorm with downbursts of high winds which passed over the airport. The aircraft broke loose from its tie downs and was flipped, damaging the port wing and the elevator. It was totaled by the insurance company and was sold sometime later.

The Club acquired another Cessna C-172, N733VS for their use.

CHAPTER 11

THE CLEMSON LADY FLYING TIGERS AND OTHER EAGLES
2000 - 2006

And it came to pass as full-time female students were admitted to Clemson that some of the students decided they wanted to take flying lessons so they joined the Clemson Flying Club. One of the first females to get their private pilot certificate was Caroline Burley Miller. She also got her commercial, multi-engine, and instrument ratings and was later employed for a number of years by the Clemson Athletic Department to fly their Aero Commander Aircraft for a number of years. Over the years there have been a number of Lady Flying Tigers and some of them are as follows:

1. Lois Nice:

Very active member, pilot, and mentor for the young ladies during the 90's. Her instructor was Randy Collins.

2. Cookie Fitch:

One of the older members of the flying club who joined in 2001 and has her own aircraft, a Cessna 172. She was 57 years old when she joined the club. Her instructor was Randy Collins.

While at Clemson she was graduated with a BS in Architecture and later started work toward a masters degree but decided to pursue a degree in Construction Science and Management.

She graduated in 1981. She worked in the construction field (Commercial and residential) for several years and then went into business for herself offering consulting services for about 8 years. She later began working with a Developer who developed small shopping centers. During the next few years they developed 12 CVS stores.

She flew many hours as passenger in a Beech Baron to these construction sites.

During this time she decided that she really wanted to learn how to fly just in case she had to land the aircraft if the pilot just happened to have an emergency and would be unable to fly the Barron. So she began to prepare herself by purchasing a Private Pilots VCR program from Sporty's Pilots Shop during Christmas of 2000. This was a Christmas present from her husband who really wasn't enthusiastic but he decided that if she really wanted to learn to fly he would be all for it. So in 2001 she contacted the FBO at the Clemson Airport to determine if they offered flight training. Since she was a Clemson alumnus she was told that she could join the Clemson University Flying Club which she did. She started in the club's Cessna 172 but she later switched to the club's Cessna 150 as suggestedby her instructor Randy Collins. The smaller aircraft was easier tofly and less expensive than the larger C-172 in which she took her introduction to flight training in.

She completed her private pilot training and received her Private Pilot license on February 12, 2002. Later she purchased her own airplane N2650L from Randy and she has flown a number of cross country flights with her husband Al who acted as the navigator to Arkansas, Florida, and other weekend locations. She purchased her aircraft N2650L from Randy on the condition that he would teach Al to at least learn how to land the aircraft. Al was soloed but didn't continue his flying lessons. During one of their flights back from Arkansas they had to deviate for thunderstorms and during that flight they had an in-flight emergency but Al picked out a nearby field and landed safely. Her instructor had taught her to respect the weather and the terrain when flying and it paid off in that flight. She is now working on her instrument rating.

One big thrill she had recently was doing three take-off and landings with the owner/instructor in a Douglas DC-3 while flying it from the left seat. It was the first time she had flown a tail dragger. She says that she wouldn't take anything for the experiences and the community of friends that she gained from her membership in the club.

3. Janis Miller:

She got her Private Certificate in 1994. Her Instructor was Terry Leap.

4. Janelle Lipscome:

She got her Private Certificate after 99 days while a member of the club in 2001. Her Instructor was Terry Leap.

5. Jennifer Grower:

She got her Private Certificate in 2003. Her Instructor was Terry Leap.

6. Nancy Johnson:

She was a member in 1976. She was probably the first female to be a member of the club and receive her Private Pilots Certificate while a member of the Club.

7. Susan Pagent:

She was the second female to get her Private Pilots Certificate while a member of the Club.

8. Caroline Burley Miller:

She was the third female in the club to solo. She later get her Private Certificate, her multi-engine and Instrument ratings and later became a pilot for the Clemson Athletic Association, flying their Aero Commander Aircraft.

She joined the Aero Club in 1982, the summer of her sophomore year. Her instructor was Heyward Douglas. She later received he Private Pilot Certificate in 1982. She was graduated with a degree in Parks, Recreation, and Tourism Management and then went to work in Columbia, SC.

After a short while she decided to work toward her multi-engine, instrument, and commercial ticket, obtaining them in July, 1988. During her instrument training she was able to fly quite a bit in the right seat of Clemson's Cessna 421 and later she was hired by Clemson's Athletic Department as a co-pilot for the Cessna 421.

Later Clemson purchased an Aero Commander in October 1988 and she became the co-pilot for the Presidents Office aircraft. She went on to become Captain and flew for 10 years in that position. After the birth of her second child she resigned from her position. After her third child she began flying again and is a part time pilot for the Athletic Department's Aero Commander.

She says she owes a lot to Clemson University and the Clemson University Flying Club for the excellent training she received.

9. Kim Cobb Gramling:

She was a member during 1991 and received her Private Pilots Certificate

10. Ellen Miller:

She was a member during 1991 also and received her Private Pilots Certificate.

11. Vonda Brandt:

She was a member in 1998 and received her Private Pilots Certificate.

12. Ellen Bether:

She was a member in 1998 and received her Private Pilots Certificate.

13. Vicky King:

She was a member in 1991 and received her Private Pilots Certificate.

14. Stacy Yikes: Major USAF

She came to Clemson via an AFROTC scholarship and her major studies were in Mechanical Engineering, graduating from Clemson University in 1989. As a freshman she was a member of the AFROTC, Arnold Air Society and the Tiger Staff. She was a member of the Tiger Band her Senior year. While attending Clemson she was a member of the Clemson Flying Association, soloed, and later obtained in 1989 her Private Pilot Certificate. Her flight instructor was George Rowland. Upon graduation she was commissioned a 2nd Lt USAFR and entered the USAF working on Satellite Systems and the USAF Space Program. She later received her Masters Degree in Aero Engineering from California State Polytechnic University. Next she served in the National Reconnaissance Office in several leadership positions.

She later received another Masters in Strategic Intelligence from the Joint Military College of the Defense Agency. She studied international relations while in Portugal and was later an Air Staff Congressional Liaison. Her present USAF job is as a political military planner for the Joint Chiefs of Staff in the Pentagon. She maintains bilateral military relationships with several European allies.

15. Joan Lazowski:

She was a member during 1983 and received her Private Pilots certificate.

There are probably other females that were members of the club that received their private pilots license or other ratings and possibly a few that never finished their training but the author has not been able to locate them.

The other Eagles

Dr. Doyce Graham, a retired professor of Crop and Soil and Environmental Science at Clemson University was one of the Old Eagles that the author contacted during research.

He joined the Club in April 2001 at the learning age of 62. The rental rate for the club planes was a real factor in his joining and training with CUFC planes. In fact his group was one of the last to be offered a block rate of 20 hours that was most attractive as to price/hour.

He soloed June 30, 2001 at Pickens County airport (LQK) and completed his private. The Club had moved to Pickens County, in early June due to a runway resurfacing project at Clemson Airport (CEU). That meant a drive of about 20 minutes instead of 5 minutes. His instructor was Shane Valentin. Shane did not instruct full time but their schedules were not a hindrance to scheduling airplanes and times for Doyce's lessons. Doyce noted that the adoption of a computer scheduling of the club planes was a real improvement over the paper signup being used when he first started his training.

He has continued to fly and purchased an RV6A, N567RV. Aircraft is a homebuilt with a160 hp engine, constant speed prop, day/night VFR. His primary flying is cross country to visit family. He and his wife routinely fly to Minnesota and Texas.

At least three of his fellow students have purchased planes after finishing training with the club. Butch Jones has a Cessna 182, Cookie Fitch has Cessna 172, and Dan Schmidt has a Luscombe 8A that is 49 years old in June of this year. Dan and Doyce share a hangar at CEU.

Doyce now has 490 plus hours. He has been flying just over 100 hours/year for past two years. Doyce routinely attends CUFC club meetings and has made all of the spring cookouts.

CHAPTER 12

OCONEE COUNTY REGIONAL AIRPORT

Today, as this is written in 2006, the Oconee Regional Airport is a jewel in Oconee County. The airport now serves 123 businesses in and around the surrounding area. Approximately 60 percent of the air traffic is industrial, corporate, and freight, the remainder being recreational. The airport Terminal was offically opened in 1998 and the airport management and its operation was under the leadership of Marion Lyles. Kevin Short is now the Airport Director. The airport has an airport advisory board and there are six members at present, these members are: Chairman Wayne Rholletta, Vice Chair Heinz Rost, Thomas Luke, Fred Golden, Alvin Rochester, and Robert Edwards. The Aircraft Maintenance Services are provided by a contractor VansSurdam,
Vintage Aircraft.

There are 73 aircraft based at the airport and they are as follows:
61 single engine aircraft (General Aviation)
5 multi engine aircraft (Transient General Aviation)
2 Jet Aircraft (Air Taxi)
2 Helicopter (Military)
1 Glider
2 Ultra-lights

Clemson University itself operates their two Aero Commanders, multi-engine aircraft, from their own facilities on the east end of the airport,. There is also a Sport Parachute Club called the Flying Tigers Sport Parachute Club that operates there at the airport. This parachute club is one of the oldest parachute clubs in the Southeast.

The Clemson Aero Club of the past is now the Clemson University Flying Club and is one of the oldest continious organizations on the Clemson University Campus. The Club for the year 2005-2006 has 15 members and the club officers are President, Brandon Arrington, Vice President, Jimmy Leesing, Treasurer, Brian Corbett, Secretary, David Kuhlman, Safety Officer, David Kuhlman, and Faculty Advisor, Heyward Douglas. The Club has two aircraft for their use, a Cessna C-152, N65407 and a Cessna C-172, n 733VS. There are two part time flight Instructors from the Clemson faculty, Randy Collins and Terry Leap. Other part time instructors are David Rood, Shane Valentine and Brandon Arrington.

The club meets on the first Tuesday of each month during the fall school session and during the summer session. Rates for rental of the aircraft are $49.00 per hour for the C-152 and $62.00 per hour for the C-172. Flight Instruction is $25 to $35 per hour. Dues are $50.00 per semester. To join the club there is a one time deposit fee of $35.00 and it is returned to the member after graduation and there is a $5.00 processing fee.

CHAPTER 13

AVIATION AND SPACE PIONEERS

Although I have tried to research and compile as much of the Clemson Aviation and Space Heritage as available. I decided to include a bio listing of those Clemson Graduates that I was able to find that have distinguished themselves in Aviation and Space. In some cases there is some biographical information that is found in the previous chapters but more information is furnished here.

GENERAL MARK EDWARD BRADLEY

Retired July 31, 1965. Died May 22, 1999

Gen. Mark Edward Bradley is commander of the U.S. Air Force Logistics Command, which has the primary mission of keeping every weapon system in the Air Force inventory operational at the least possible cost.

General Bradley was born in Clemson, S.C., in 1907. After three years at Clemson High School, he attended Clemson A&M College and in July 1926 entered the U.S. Military Academy, graduated in June 1930, and was commissioned a second lieutenant of Field Artillery. He graduated from flying school in 1931 and was rated a pilot.

General Bradley's first assignment as an Air Corps officer was atSelfridge Field, Mich., where he served as a fighter pilot in the 27th Pursuit Squadron. In 1934 he graduated from the Air Corps Technical School and was assigned to the 18th Pursuit Group at Schofield Barracks, Hawaii.

Graduating from the Engineering School at Wright Field, Ohio, in July 1938, General Bradley was assigned to the Flight Test Section as a test pilot. Then he was, in succession, P-47 project officer, a project officer in the Pursuit Branch of the Materiel Division, chief of the Fighter Branch, and chief of the Flight Test Section, where he was responsible for a key modification to the P-51 Mustang. He came up with the idea of putting a fuel tank behind the pilot's seat (80 to 100 gal) to increase the P-51B Mustang Fighter's range to allow the fighters to escort the bombers to Berlin. Retro-fit kits were provided to the fighter in early 1943.

Going to Europe in January 1945, General Bradley became deputy commanding officer of the 1st Tactical Air Force Service Command. In May 1945 he was assigned to the Fifth Air Force in the Philippines, becoming chief of staff in October 1945, on its subsequent move to Japan. In February 1946 he assumed command of the 301st Fighter Wing on Okinawa where he remained until October 1946.

In November 1946 General Bradley was assigned to the Armed Forces Staff College at Norfolk, Va., as assistant director of the college's Plans and Operations Division.

Returning to Wright-Patterson Air Force Base, Ohio, in April 1948, the general was appointed to the Air Materiel Command's Directorate of Procurement and Industrial Planning as deputy director, becoming director of procurement and production in July

1951.

In April 1953, General Bradley was named assistant chief of staff formateriel, U.S. Air Forces in Europe, with headquarters in Wiesbaden, Germany. In August he became deputy commander in chief and chief of staff, USAFE.

Returning to the United States in July 1956, General Bradley was appointed assistant deputy chief of staff for materiel at Air Force Headquarters, Washington, D.C. He became deputy chief of staff for materiel on June 30, 1959. His position was redesignated deputy chief of staff for systems and logistics in 1961.

On July 1, 1962, General Bradley became commander, Air Force Logistics Command at Wright-Patterson Air Force Base, Ohio, with concurrent promotion to general.

Among his decorations are the Distinguished Service Medal, the Legion of Merit, the Bronze Star Medal with oak leaf cluster, the Air Medal and the French Croix de Guerre. Now a command pilot, he flew six combat missions during World War II for a total of 32 combat hours. His campaign ribbons - all with battle stars - include the European-African-Middle Eastern Campaign Medal, Asiatic-Pacific Campaign Medal and the Philippine Liberation Ribbon.

GENERAL THOMAS S. MOORMAN JR

Retired Aug. 1, 1997

General Thomas S. Moorman Jr. is vice chief of staff, Headquarters U.S. Air Force, Washington, D.C.

General Moorman was born in Washington, D.C. He was commissioned through the Air Force Reserve Officer Training Corps program as a distinguished military graduate in 1962. The general has served in a variety of intelligence and reconnaissance related positions within the United States and worldwide. While stationed at Peterson Air Force Base, Colo., in 1982, he became deeply involved in the planning and organizing for the establishment of Air Force Space Command. During his Pentagon tour in 1987, he also provided program management direction for development and procurement of Air Force surveillance, communications, navigation and weather satellites, space launch vehicles, anti-satellite weapons and ground-based and airborne strategic radars, communications and command centers. He additionally represented the Air Force in the Strategic Defense Initiative program and was authorized to accept SDI program execution responsibilities on behalf of the Air Force. As commander and vice commander of Air Force Space Command, General Moorman was responsible for operating military space systems, ground-based radars and missile warning satellites, the nation's space launch centers at Patrick Air Force Base, Fla., and Vandenberg Air Force Base, Calif., the worldwide network of space surveillance radars, as well as maintaining the Intercontinental Ballistic Missile (ICBM) force.

EDUCATION

1962 Bachelor's degree in history and political science, Dartmouth College, Hanover, N.H.

1965 Squadron Officer School, Maxwell Air Force Base, Ala.

1972 Master's degree in business administration, Western New England College, Springfield, Mass.

1975 Air Command and Staff College, Maxwell Air Force Base, Ala.

1975 Master's degree in political science, Auburn University, Ala.

1979 Air War College, by seminar

1980 National War College, Fort Lesley J. McNair, Washington, D.C.

ASSIGNMENTS

1. July 1962 - August 1965, intelligence officer, B-47 bombardment wing, Schilling Air Force Base, Kan.
2. August 1965 - October 1966, mission planner, 9th Strategic Reconnaissance Wing, Beale Air Force Base, Calif.
3. October 1966 - November 1967, operations officer, 432nd Reconnaissance Technical Squadron, Udorn Royal Thai Air Force Base, Thailand
4. November 1967 - November 1970, reconnaissance intelligence staff officer, 497th Reconnaissance Technical Group, Schierstein, West Germany
5. November 1970 - August 1975, assistant director of evaluation, later executive officer, Air Force Special Projects Production Facility, Westover Air Force Base, Mass.
6. August 1975 - August 1979, executive, later deputy director of plans and programs, Office of Space Systems, Office of the Secretary of the Air Force, Washington, D.C.
7. August 1979 - June 1980, National War College, Fort Lesley J. McNair, Washington, D.C.
8. June 1980 - August 1981, deputy military assistant to the secretary of the Air Force, Washington, D.C.
9. August 1981 - March 1982, director of space operations, North American Aerospace Defense Command, Cheyenne Mountain Complex, Colo.
10. March 1982 - August 1982, deputy director, Space Defense, Office of the Deputy Chief of Staff for Plans, Peterson Air Force Base, Colo.
11. August 1982 - July 1984, first director, commander's group, Air Force Space Command, Peterson Air Force Base, Colo.
12. July 1984 - March 1985, vice commander, 1st Space Wing, Peterson Air Force Base, Colo.
13. March 1985 - October 1987, director of space systems, Office of the Secretary of the Air Force, Washington, D.C.
14. October 1987 - March 1990, director of Space and Strategic Defense Initiative programs, Office of the Assistant Secretary of the Air Force for Acquisitions, the Pentagon, Washington, D.C.
15. October 1987 - March 1990, special assistant for Strategic Defense Initiative to the vice commander of Air Force Systems Command, Andrews Air Force Base, Md.
16. March 1990 - March 1992, commander, Air Force Space Command, Peterson Air Force Base, Colo.
17. March 1992 - July 1994, vice commander, Air Force Space Command, Peterson Air Force Base, Colo.
18. July 1994 - present, vice chief of staff, Headquarters U.S. Air Force, Washington, D.C.

MAJOR AWARDS AND DECORATIONS

Distinguished Service Medal with oak leaf cluster
Defense Superior Service Medal
Legion of Merit with oak leaf cluster
Meritorious Service Medal with oak leaf cluster
Air Force Commendation Medal with oak leaf cluster
National Intelligence Distinguished Service Medal

OTHER ACHIEVEMENTS

1991 National Geographic Society's Gen. Thomas D. White U.S. Air Force Space Trophy, awarded to the individual who has made the most outstanding contribution to the nation's progress in space

1993 Eugene M. Zuckert Management Award, which yearly recognizes outstanding management achievements by a top-level Air Force manager

1994 Ira C. Eaker Fellowship Award, sponsored by the Air Force Association Aerospace Education Foundation to honor significant accomplishments in Air Force space systems

1995 Dr. Robert H. Goddard Memorial Trophy, the premier award of the National Space Club presented annually for significant contribution to the country's progress in the field of rocketry and astronautics

EFFECTIVE DATES OF PROMOTION

Second Lieutenant Jul 10, 1962
First Lieutenant Jun 9, 1965
Captain Jan 2, 1967
Major May 1, 1973
Lieutenant Colonel Nov 1, 1978
Colonel Jul 1, 1981
Brigadier General Apr 1, 1985
Major General Feb 1, 1988
Lieutenant General Apr 1, 1990
General Aug 1, 1994

MISSLE AND SPACE INFORMATION

Master Space Badge
(Current as of April 1995)

Added note from author:

General Mooreman, although not a Clemson graduate, spent almost ever summer in the 50's, at Clemson visiting his grandmother and his other relatives there at Clemson. The Clemson spirit has been a member of their family for about over 110 years.

LT GENERAL THOMAS S. MOORMAN

Retired Aug. 1, 1970. Died Dec. 23, 1997

Lieutenant General Thomas Samuel Moorman Jr. is the superintendent of the Air Force Academy at Colorado Springs, Colorado. The Air Force Academy provides instruction, experience and motivation to each cadet so that he will graduate with the knowledge, character and qualities needed for leadership that are essential to his progressive development as a career officer in the U.S. Air Force.

General Moorman was born at the Presidio of Monterey, Calif., in 1910. He attended John J. Phillips High School in Birmingham, Ala. In 1933 he graduated from the U.S. Military Academy at West Point, N.Y., and entered the Air Corps flying training school at Randolph Field, Texas.

He earned his pilot wings in October 1934 and was assigned to the 4th Observation Squadron, 5th Composite Group at Luke Field, Hawaii. In July 1936, General Moorman was reassigned to the 97th Reconnaissance Squadron at Mitchel Field, N.Y. for a 12-month tour.

In 1937 General Moorman entered the field of meteorology, obtained a master of science degree from the California Institute of Technology and embarked on a sphere of activity which would dominate the next 20 years of his career. After a two-year assignment as Weather Officer at Randolph Field, Texas, he gained valuable information in meteorology through further study at the Massachusetts Institute of Technology.

In July 1941,General Moorman was assigned to Air Corps Headquarters in Washington, D.C., where he served as chief climatologist and assistant director of the Air Corps Research Center and liaison officer to the U.S. Weather Bureau. From May 1943 until December 1943, General Moorman was commanding officer of the 21st Weather Squadron, first at Bradley Field, Conn., and later in England.

General Moorman became staff weather officer and later director of weather for Ninth Air Force in the European Theater of Operations, a position he held from January 1944 until December 1944. From January 1945 until June 1945, he served as commanding officer of the 21st Weather Squadron and Staff Weather Officer for Ninth Air Force.

After V-J Day in 1945, General Moorman returned to the U.S. as deputy chief of staff for Air Weather Service. One year later he became air weather officer at Headquarters Army Air Forces and remained in that position until he entered the Air War College at Maxwell Air Force Base, Ala.

From July 1948 until June 1951, General Moorman served with the Far East Air Forces in Tokyo as commander of the 2143d Air Weather Wing (now the 1st Weather Wing), and upon his return to the United States he became deputy commander of the Air Weather Service at Andrews Air Force Base, Md.

On April 22, 1954, his 16 years of weather operations culminated in his appointment as Air Weather Service commander, responsible for providing environmental support to the Air Force and Army through weather central forecast centers, base weather stations, observation sites, world-wide weather reconnaissance and atmospheric sampling.

Returning to the Far East in 1958, General Moorman assumed command of the Thirteenth Air Force at Clark Air Base, Philippines. He was responsible for air operations throughout Southeast Asia, as well as for the air defense of the Philippines in conjunction with the Philippine Air Force.

On July 28, 1961, General Moorman became vice commander in chief, Headquarters Pacific Air Forces, Hickam Air Force Base, Hawaii, and on July 1, 1965, General Moorman became superintendent of the Air Force Academy at Colorado Springs, Colo.

Among his awards and decorations are the Distinguished Service Medal, Legion of Merit with two oak leaf clusters, Bronze Star Medal and the Air Medal.
(Current as of Dec. 9, 1965)

Added Note by the Author:

Lt General Mooreman's father Thomas Samuel Mooreman was a member of the first graduating class of Clemson College in 1893. He was a career Army officer and he was assigned to Clemson College as the Commandant of Cadets in 1934 for a short time.

After his retirement, General Mooreman served as the President of Friends of the Air Force Academy Library, a trustee of the Falcon Foundation, and a member of the board of directors of the Air Academy Bank, as well as serving on other corporate boards. He was also very active in environmental issues, serving on the Colorado Air Resources Board and as an Advisor to the National Oceanic and Atmospheric Administration.

GENERAL H.T. JOHNSON

Retired August 31, 1992

General Hansford T. Johnson is commander in chief of the U.S. Transportation Command and Air Mobility Command, Scott Air Force Base, Ill.

General Johnson was born in 1936, in Aiken, S.C. He graduated from high school in 1953 and then attended Clemson College. He was the outstanding graduate in thermodynamics and aerodynamics from the U.S. Air Force Academy in 1959, received a master's degree in aeronautics from Stanford University In 1967 and a master's degree in business from the University of Colorado in 1970. The general completed Squadron Officer School in 1965, Army Command and General Staff College in 1972, National War College in 1976, and the advanced management program at Dartmouth College in 1980.

He earned his navigator wings while a cadet at the academy. Upon graduation, he attended flying training at Bartow Air Base, Fla., and then Laredo Air Force Base, Texas, receiving his pilot wings in July 1960. His first operational assignment was to the 317th Troop Carrier Wing, Evreux-Fauville Air Base, France, where he flew C-130 transports throughout Europe, Africa, the Middle East and West Asia. He continued flying with the 317th when it transferred to Lockbourne Air Force Base, Ohio.

After completing graduate school at Stanford University, General Johnson volunteered for duty in South Vietnam. In 1967 he flew as a forward air controller in support of the South Vietnamese army and U.S. Marine Corps forces in the northern province and the demilitarized zone. He directed tactical close air support strike missions against enemy supply, storage, vehicle and troop targets. During the latter half of the year, he operated out of Plelku Air Base, along the Ho Chi Min trail, at night. During his tour, he flew 423 combat missions, 71 of which were over North Vietnam or the demilitarized zone.

From May 1968 to July 1971 General Johnson was an instructor and assistant professor of aeronautics at the U.S. Air Force Academy. After graduating from Army Command and General Staff College, he served three years in the Plans Directorate, Headquarters U.S. Air Force, Washington, D.C. During his first year he was in charge of developing the tactical airlift force structure, then became leader of the team that developed and evaluated Air Force inputs to military strategy.

The general graduated from National War College in June 1976 and then was assigned to the 93rd Bombardment Wing, Castle Air Force Base, Calif., as assistant deputy commander for operations. He also performed duties as assistant deputy commander for maintenance, deputy commander for operations and vice commander. He was commander of the 22nd Bombardment Wing at March Air Force Base, Calif., from April 1979 until February 1981. General Johnson then was assigned to Headquarters Strategic Air Command, Offutt Air Force Base, Neb., as assistant deputy chief of staff for plans.

In November 1982 the general returned to Air Force headquarters as deputy director of programs and evaluation, Office of the Deputy Chief of Staff, Plans and Resources, and was chairman of the Program Review Committee. He was director of programs and evaluation, and chairman of the Air Staff Board, from July 1984 to October 1985, when he became deputy chief of staff for operations at SAC headquarters. In December 1986 he was assigned as vice commander in chief, Pacific Air Forces, Hickam Air Force Base, Hawaii. In August 1987 General Johnson became deputy commander in chief of the U.S. Central Command, MacDill Air Force Base, Fla., during Earnest Will, the United States tanker reflagging operations in the Persian Gulf. He became director of the Joint Staff, Joint Chiefs of Staff, the Pentagon, Washington, D.C., in November 1988. Effective June 1, 1992, General Johnson became the first commander of Air Mobility Command while remaining as commander in chief of the U.S. Transportation Command, a position he assumed in September 1989.

The general is a command pilot with more than 7,800 flying hours, more than 1,000 of which were flown under combat conditions. He is also a qualified navigator and parachutist. His military awards and decorations include the Defense Distinguished Service Medal with oak leaf cluster, Distinguished Service Medal, Silver Star, Legion of Merit with two oak leaf clusters, Distinguished Flying Cross with two oak leaf clusters, Defense Meritorious Service Medal, Meritorious Service Medal, Air Medal with 22 oak leaf clusters, Presidential Unit Citation, Navy-Marine Corps Presidential Unit Citation,

Joint Meritorious Unit Award, Air Force Outstanding Unit Award with "V" device and three oak leaf clusters, Air Force Organizational Excellence Award, Combat Readiness Medal, Armed Forces Expeditionary Medal with service star, Vietnam Service Medal with three service stars, Humanitarian Service Medal, Air Force Overseas Ribbon-Short, Air Force Overseas Ribbon-Long with oak leaf cluster, Air Force Longevity Service Award Ribbon with seven oak leaf clusters, Air Force Training Ribbon, Republic of Vietnam Gallantry Cross with Palm, Republic of Vietnam Armed Forces Honor Medal First Class with service star and Republic of Vietnam Campaign Medal. He is a registered professional engineer in Colorado.

He was promoted to general Oct. 1, 1989, with same date of rank.
(Current as of June 1992)

LT GENERAL GRADY L. PATTERSON
Retired

Grady L Patterson Jr was born in Calhoun Falls, SC in 1924 and after graduating from high school he attended Clemson A & M College in 1942 but was there for only one semester, leaving to enlist in the US Army Air Corps as an aviation cadet. He was completed his Primary, Basic, and Advance training and was awarded his silver wings and a 2nd Lt commission. He then continued his combat training, becoming a fighter pilot and was assigned to the Pacific War area. Here he escourted bombers to their targets in the Pacific theater of operations including raids on the Islands of Japan.

When he returned to South Carolina from the Pacific, he enrolled at the University of South Carolina. Upon graduation Grady decided to enter the USC School of Law. He earned his degree fully expecting to return to Abbeville County to practice law upon graduation. During this time he also joined the SC Air National Guard and remained with the guard retiring as a Lt General. During his service with the SC Air National Guard he was called to active duty during the Korean War and in the Berlin Crisis when the National Guards of the USA were federalized. He was awarded the Distinguished Service Cross while as a Major General in 1984.

South Carolina Attorney General Dan McLeod soon recruited Grady from law school to become a member of his staff as Assistant Attorney General. For seven years, he served the state as Assistant Attorney General traveling South Carolina. Grady spent much of his time on Interstate construction litigation, but also worked on a wide variety of other cases battling legendary court figures such as Sol Blatt and Tom Pope. Grady was also assigned as legal counsel to the State Treasurer's Office and the State Board of Bank Control (now the Board of Financial Institutions) and learned of the challenges facing rural South Carolina in the 1950's and 60's.

In 1966, four term South Carolina State Treasurer Jefferson B. Bates unexpectedly passed away and state leaders came to Grady L. Patterson, Jr., and asked him to run for the vacant post. With strong background knowledge of county and state funding practices through his years as Assistant Attorney General, Grady worked night and day, and was elected for the first time that year. Grady L. Patterson, Jr. has spent his life in the service of his state and his country. His dedication has brought accolades from voters, citizens, state and national leaders. He continued to energetically safeguard South Carolina's finances until his death in December, 2009.

MAJOR GENERAL OTIS C. MOORE

Retired July 1, 1977

General Moore was born in Charlotte, N.C., in 1926. He graduated from high school in Sumter, S.C., attended Clemson College, S.C., and Worcester Polytechnic Institute, Worcester, Mass. After graduation from the U.S. Military Academy, West Point, N.Y., in 1948 with a bachelor of science degree, he was commissioned as a second lieutenant. Pilot training was accomplished at Randolph Air Force Base, Texas, and Barksdale Air Force Base, La., in 1949.

After graduation from pilot training, he was assigned to the 43d Bombardment Wing, Davis-Monthan Air Force Base, Ariz., as a B-50 aircraft commander. In July 1954, after completing triple-rated training as a navigator-bombardier-radar operator, he was again assigned to the 43d Bombardment Wing, this time as a B-47 aircraft commander; in July 1956 General Moore was appointed chief, Standardization Division; and in October 1959 became chief of air training for the wing.

General Moore was transferred to Headquarters Fifteenth Air Force, SAC, March Air Force Base, Calif., in April 1960, where he served in the Directorate of Operations as planner for current operations including bomber deployments and airborne alert missions.

In October 1962 he was transferred to the Directorate of Plans, Headquarters Strategic Air Command, Offutt Air Force Base, Neb., where he developed operational concepts for the SR-71 aircraft, other reconnaissance systems, and the Defense Meteorological Satellite System. In 1965 he assumed the position of chief, Space Branch, Future Systems Division. During this period, he earned a master's degree from the University of Omaha.

He graduated from the National War College in June 1967 and was assigned to the Directorate of Doctrine, Concepts and Objectives, Deputy Chief of Staff, Plans and Operations, Headquarters U.S. Air Force, Washington, D.C. During this period he was director of a study entitled "Strategic Options in the 70s (STRAT 70)," which developed the concept of TRIAD and introduced the concept of limited nuclear options within the U.S. Air Force. In June 1969 he was assigned as executive officer to the chief of staff, U.S. Air Force.

In January 1971 General Moore became chief of staff, Seventh Air Force, in the Republic of Vietnam. While stationed in Vietnam, General Moore flew 25 combat missions in the F-4 Phantom jet, and 83 combat support missions. He went to Ent Air Force Base, Colo., in September 1972 and assumed command of the Fourteenth Aerospace Force, ADC. While serving as commander, his unit was awarded a Presidential Management Improvement Award.

General Moore returned to Headquarters U.S. Air Force in May 1974 as director of operations in the Office of the Deputy Chief of Staff, Plans and Operations. His additional duty in this assignment was Air Force member, Permanent Joint Board on Defense, Canada, United States. In August 1975, General Moore assumed duty as the assistant deputy chief of staff, Plans and Operations.

His military decorations and awards include the Distinguished Service Medal with oak leaf cluster, Legion of Merit with oak leaf cluster, Air Medal with oak leaf cluster, Air Force Commendation Medal with two oak leaf clusters, and the National Order of Vietnam, Fifth Class. He is a rated navigator and command pilot.

He was promoted to the grade of major general effective Nov. 1, 1972, with date of rank Aug. 1, 1969.

(Current as of Sept. 15, 1975)

MAJOR GENERAL SCOTT C. BERGREN

Retired Sep. 1, 2003

Maj. Gen. Scott C. Bergren is Commander, Ogden Air Logistics Center, Hill Air Force Base, Utah. The center manages $9.6 billion through a work force of more than 12,000 people. It provides worldwide logistics management, engineering, supply, contracting and depot maintenance for the F-16, A-10, KC-135 and C-130, the Minuteman and Peacekeeper intercontinental ballistic missiles, space and C3I programs, and 140,000 Air Force and Department of Defense commodities. General Bergren was born in Mineola, N.Y. He entered the Air Force and was commissioned in 1970 as a distinguished graduate of the Air Force ROTC program at Clemson University. He has commanded a training wing, aircraft maintenance squadron, supply squadron and support group, and served as deputy commander for maintenance of a fighter wing. Prior to his current assignment, he served as Director of Maintenance for the Deputy Chief of Staff for Installations and Logistics at Headquarters
U.S. Air Force.

EDUCATION

1970 Bachelor of arts degree in economics, Clemson University, Clemson, S.C.
1974 Squadron Officer School, Maxwell Air Force Base, Ala.
1981 Master's degree in political science, Auburn University, Auburn, Ala.
1981 Air Command and Staff College, Maxwell Air Force Base, Ala.
1990 Air War College, Maxwell Air Force Base, Ala.
1996 Program for Senior Managers in Government, John F. Kennedy School of Government, Harvard University, Cambridge, Mass.

ASSIGNMENTS

1. May 1970 - March 1971, student, undergraduate navigator training course, Mather Air Force Base, Calif.
2. March 1971 - October 1971, F-4 Phantom combat crew training, George Air Force Base, Calif.
3. October 1971 - October 1972, F-4 Phantom weapons system officer, 8th Tactical Fighter Wing, Ubon Royal Thai Air Force Base, Thailand
4. October 1972 - July 1976, F-4 Phantom instructor, weapons system officer and flight examiner, 1st Tactical Fighter Wing, MacDill Air Force Base, Fla.
5. July 1976 - July 1977, Air Staff operations officer and assistant executive officer to the Deputy Chief of Staff for Programs and Resources, Air Staff Training Program, Headquarters U.S. Air Force, Washington, D.C.
6. July 1977 - July 1980, instructor, navigator and exchange officer, 37th Operational Conversion Unit, Royal Air Force Honington, England
7. July 1980 - July 1981, student, Air Command and Staff College, Maxwell Air Force Base, Ala.
8. July 1981 - October 1984, Director of Operations, Force Analysis Division, Headquarters Tactical Air Command, Langley Air Force Base, Va.
9. October 1984 - July 1985, Special Assistant to the Commander, Headquarters Tactical Air Command, Langley Air Force Base, Va.
10. July 1985 - June 1987, Commander, 325th Aircraft Generation Squadron, and later, Assistant Deputy Commander for Maintenance, 325th Tactical Training Wing, Tyndall Air Force Base, Fla.
11. June 1987 - June 1989, Deputy Commander for Maintenance, 33rd Tactical

Fighter Wing, Eglin Air Force Base, Fla.
12. June 1989 - June 1990, student, Air War College, Maxwell Air Force Base, Ala.
13. June 1990 - July 1991, Air University Chair for the Chief of Staff of the Air Force, Maxwell Air Force Base, Ala.
14. July 1991 - November 1991, Commander, 554th Supply Squadron, Nellis Air Force Base, Nev.
15. November 1991 - April 1993, Commander, 554th Support Group, Nellis Air Force Base, Nev.
16. April 1993 - August 1994, Deputy Director for Logistics and Security Assistance, U.S. Central Command, MacDill Air Force Base, Fla.
17. August 1994 - March 1996, Executive Officer to the Commander in Chief, U.S. Central Command, MacDill Air Force Base, Fla.
18. March 1996 - February 1997, Vice Commander, San Antonio Air Logistics Center, Kelly Air Force Base, Texas
19. February 1997 - April 1999, Commander, 82nd Training Wing, Sheppard Air Force Base, Texas
20. April 1999 - January 2000, Director of Maintenance, Deputy Chief of Staff for Installations and Logistics, Headquarters U.S. Air Force, Washington, D.C.
21. January 2000 - present, Commander, Ogden Air Logistics Center, Hill Air Force Base, Utah

FLIGHT INFORMATION
Rating: Senior navigator
Flight hours: 1,622
Aircraft flown: F-4 Phantom, Royal Air Force Buccaneer aircraft

MAJOR AWARDS AND DECORATIONS
Distinguished Service Medal
Silver Star
Defense Superior Service Medal
Legion of Merit
Distinguished Flying Cross with five oak leaf clusters
Purple Heart
Meritorious Service Medal with five oak leaf clusters
Air Medal with 16 oak leaf clusters
Air Force Commendation Medal
Air Force Achievement Medal
Combat Readiness Medal
National Defense Service Medal with service star
Vietnam Service Medal with three service stars
Air Force Overseas Ribbon-Long
Air Force Longevity Service Award Ribbon with five oak leaf clusters
Small Arms Expert Marksmanship Ribbon
Republic of Vietnam Gallantry Cross with Palm
Republic of Vietnam Campaign Medal

Effective dates of Promotion
Second Lieutenant June 3, 1970
First Lieutenant Nov 14, 1971
Captain Oct 14, 1973
Major Dec 4, 1978

Lieutenant Col Feb 1, 1984
Col Dec 1, 1996
Brigadier General June 1, 1996
Major General July 1, 1999
Current as of May 2002

MAJOR GENERAL JONES E. BOLT

Retired Aug. 1, 1973

Maj. Gen. Jones E. Bolt is inspector general, Headquarters Air Training Command, Randolph Air Force Base, Texas.

General Bolt was born at Ware Shoals, S.C., in 1921. He graduated from high school at Williamston, S.C., and entered Clemson University in 1938 from which he graduated in 1942 with a degree in engineering and, through the Reserve Officers' Training Corps program, a reserve commission as second lieutenant in the U.S. Army.

He entered active military duty in August 1942 and received flying training at Orangeburg, S.C., and Macon and Moultrie, Ga. He completed advanced flying training in March 1943, and was assigned as a T-6 and P-40 aircraft pilot instructor in advanced flying school at Spence Field, Ga. In March 1944 he was transferred to Florida for P-47 combat training.

In April 1944 he was assigned to the 362d Fighter Bomber Group in England. He flew 15 combat missions in the European Theater of Operations before he was forced to bail out of his airplane because of engine failure. He landed in a German Command Post just outside of Paris, France, and was taken prisoner. General Bolt was sent to Stalagluft 3, a prisoner of war camp near Sagan, Germany, and was one of the prisoners of war on the infamous march from Stalagluft 3 to Stalag 7A at Mooseburg during January and February 1945. He was liberated from the prisoner of war camp in April 1945.

Upon return to the United States in 1945 he attended Aircraft Engineering School at Chanute Field, Ill. In June 1946 he was assigned as flight test maintenance officer for F-80 aircraft at Williams Field, Ariz., and later was assigned as Deputy Commander for Maintenance, maintenance and Supply Group.

In 1948 he assisted in the formation of and flew right wing on the first Air Force jet acrobatic team, "The Acrojets." The team remained intact until August 1950 at which time it was deactivated, because of the Korean War. After reactivation in 1952 the team was renamed the Thunderbirds.

In August 1950 he was transferred to the 86th Fighter Wing in Munich, Germany, and served as commander of the 526th Fighter Bomber Squadron.

In November 1953 he returned to the United States and attended the January-June class of the Air Command and Staff College at Maxwell Air Force Base, AL. After graduation he was transferred to the War Plans Division, Directorate of Plans, Headquarters U.S. Air Force, Washington, D.C., where for the first three years he was involved in U.S. Air Force commitments and deployments to NATO and correlating war plans associated with these forces. The last year he was a member of the Force Structure Committee which involved future structure plans for the U.S. Air Force.

In July 1958 General Bolt was assigned as commander of the 53d Fighter Interceptor Group, Sioux City, Iowa, and in October 1959 he became commander of the 328th Fighter Wing, Aerospace Defense Command, at Richards-Gebaur Air Force Base, Mo. He was assigned to the 18th Tactical Fighter Wing on Okinawa as deputy commander in July 1961.

February 1963 after completing F-105 aircraft operational training at Nellis Air Force Base, Nev., General Bolt planned and led the first nonstop flight of "Thunderchiefs" from

Hickam Air Force Base, Hawaii, to Kadena Air Base, Okinawa, via Wake and Guam This marked the first time a single engine fighter had flown such long distances over water. In May 1964 he became commander of the 18th Tactical Fighter Wing. He was assigned to Headquarters Twelfth Air Force in Waco, Texas, as director of safety in August 1964. He assumed command of the 4453d Combat Crew Training Wing at Davis-Monthan Air Force Base, Ariz., in July 1965. The wing trained F-4C pilots for combat duty in Southeast Asia and transition pilots of the fighter wings converting to the F-4 aircraft.

In October 1966 General Bolt was transferred to Southeast Asia as commander of the 12th Tactical Fighter Wing, equipped with F-4s at Cam Ranh Bay, Republic of Vietnam. In March 1967 he was sent to Da Nang Air Base to command the 366th Tactical Fighter Wing. During that assignment he regularly flew missions into the Hanoi area and was awarded the Distinguished Flying Cross with oak leaf cluster in addition to the Air Medal with nine oak leaf clusters. He joined Headquarters Seventh Air Force as director of the Tactical Air Control Center in June 1967 with responsibilities for planning and execution of air operations in the southern part of North Vietnam and throughout South Vietnam. Following this assignment he was awarded the Air Force Distinguished Service Medal, the third Distinguished Flying Cross and the tenth Air Medal.

He returned to Headquarters U.S. Air Force in May 1968 as deputy director for strike forces in the Directorate of Operations, Deputy Chief of Staff, Plans and Operations. In March 1970 General Bolt became deputy chief of staff for operations, Headquarters U.S. Air Forces in Europe, with headquarters at Lindsey Air Station, Wiesbaden, Germany.

General Bolt became Air Training Command inspector general in August 1971.

His military decorations and awards include the Distinguished Service Medal, Distinguished Flying Cross with two oak leaf clusters, Air Medal with 10 oak leaf clusters, Air Force Commendation Medal with oak leaf cluster, Purple Heart, Republic of Vietnam Air Force Distinguished Service Order First Class and the Air Gallantry Cross. A command pilot, General Bolt has more than 7,000 hours of flying time and 160 combat missions in Southeast Asia.

General Bolt was promoted to the temporary grade of major general effective Feb. 24, 1970, with date of rank Aug. 10, 1965.

(Current as of Sept. 15, 1971)

MAJOR GENERAL JOHN L. MARTIN JR.

Retired Feb. 1, 1970

Major General John Landrum Martin Jr., is assistant for Systems Acquisition Management, Headquarters Air Force Systems Command, Andrews Air Force Base, Md.

General Martin was born in Spartanburg, S.C., in 1919. He graduated from Fairforest High School, S.C., studied mechanical engineering at Clemson College, S.C., from 1937 to 1940. He entered the U.S. Army in July 1940 as an aviation cadet, attended flying training at Hicks, Randolph and Kelly fields, Texas, and received his pilot wings and commission as a second lieutenant in March 1941.

He was assigned as a flying instructor at Kelly and Ellington fields, Texas, from 1941 to 1943. Following this assignment, he was stationed at Fort Worth, Texas; Tucson, Ariz.; and Great Bend, Kan., for training in B-24, B-17 and B-29 aircraft. From April 1944 until March 1945, he served with the 444th Bombardment Group in the China-Burma-India Theater of Operations as aircraft commander and squadron operations officer and flew 14 combat missions. He returned to the United States in March 1945 and served as a commander of a B-29 training group and flying training staff officer at Randolph Field, Texas.

He entered Polytechnic Institute of Brooklyn in 1946 and received a bachelor of science degree in aeronautical engineering in 1948. After graduation he was stationed at Roswell, N.M., as maintenance squadron commander and group materiel officer. From 1949 to 1951 he attended the Massachusetts Institute of Technology, Boston, Mass., and received a master of science degree in aeronautical engineering. He then was assigned as assistant professor at the Air Force Institute of Technology at Wright-Patterson Air Force Base, Ohio, in June 1951. In July 1954 he was assigned as chief of the Flight Control Laboratory, Wright Air Development Center, Ohio.

General Martin became deputy director of advanced technology at Headquarters U.S. Air Force, Washington, D.C., in July 1958. In 1960 he was assigned to the Office of the Secretary of the Air Force where he served as deputy director, Office of Missile and Satellite Systems; later as deputy director, Office of Space Systems; in 1962 became director, Office of Space Systems; and in August 1964 assumed duties as vice director of special projects at Los Angeles Air Force Station, and also served as assistant deputy commander for satellite programs, Space Systems Division, Air Force Systems Command.

In July 1965 he became director of special projects, and deputy commander for satellite programs at Space Systems Division, now the Air Force Space and Missile Systems Organization. In August 1969 General Martin was assigned to Headquarters Air Force Systems Command as assistant for systems acquisition management.

His military decorations include the Distinguished Service Medal, Legion of Merit with oak leaf cluster, Distinguished Flying Cross with oak leaf cluster, Air Medal with two oak leaf clusters and Distinguished Unit Citation Emblem. He is a command pilot.

(Current as of Nov. 1, 1969)

MAJOR GENERAL JOHN F MILLER JR

Retired Jan. 1, 2000

Maj. Gen. John F. Miller Jr. is the chief of staff, U.S. Atlantic Command, Norfolk, Va. He is the principal adviser and assistant to the commander in chief and deputy commander in chief. He is responsible for developing, reviewing and managing the command's war fighting concerns, plans and policy issues, and defense initiatives affecting the largest U.S. unified command. He provides policy and procedural guidance to the executive staff on personnel, intelligence, operations, logistics, plans and programs, communications, training and financial matters. He also supervises administrative functions within the command.

The general entered the Air Force in 1968 as a distinguished graduate of the Reserve Officer Training Corps program. He has commanded the 4450th Tactical Squadron, the first operational squadron of the F-117A stealth fighter; 8th Tactical Fighter Wing, the Wolf Pack; 49th Fighter Wing; and 57th Wing. He was selected Air Combat Command's outstanding wing commander for 1994. The general is a command pilot with more than 3,600 fighter flying hours, including 126 combat missions, 86 of which were over North Vietnam.

EDUCATION

1968 Bachelor of arts degree in history, Clemson University

1973 U.S. Air Force Fighter Weapons Instructor School, Nellis Air Force Base, Nev.

1982 Master's degree in military arts and sciences, U.S. Army Command and General Staff College, Fort Leavenworth, Kan.

1988 National War College, Fort Lesley J. McNair, Washington, D.C.

ASSIGNMENTS

1. May 1968 - May 1969, student, undergraduate pilot training, Laughlin Air Force Base, Texas

2. June 1969 - February 1970, F-105 pilot training, McConnell Air Force Base, Kan.
3. June 1970 - October 1970, F-105D pilot, 357th Tactical Fighter Squadron, Takhli Royal Thai Air Force Base, Thailand
4. October 1970 - May 1971, F-105G pilot, 6010th Wild Weasel Squadron, Korat Royal Thai Air Force Base, Thailand
5. June 1971 - July 1972, A-7D pilot, 354th Tactical Fighter Squadron,
Davis-Monthan Air Force Base, Ariz.
6. August 1972 - January 1979, A-7D instructor pilot, flight examiner and fighter weapons instructor, 23rd Tactical Fighter Wing, England Air Force Base, La.
7. March 1979 - June 1981, F-16 multinational operational test and evaluation team member, Hill Air Force Base, Utah
8. July 1981 - June 1982, student, U.S. Army Command and General Staff College, Fort Leavenworth, Kan.
9. July 1982 - January 1985, chief, Tactical Division and deputy director, Directorate of Standardization and Evaluation, Headquarters U.S. Air Forces in Europe, Ramstein Air Base, West Germany
10. June 1985 - June 1987, F-117A commander, 4450th Tactical Squadron, and deputy commander, 4461st Support Group, Tonopah Test Range, Nev.
11. August 1987 - June 1988, student, National War College, Fort Lesley J. McNair, Washington, D.C.
12. June 1988 - June 1989, chief, Tactical Forces Division, Deputy Chief of Staff for Programs and Resources, Headquarters U.S. Air Force, the Pentagon, Washington, D.C.
13. August 1989 - May 1990, deputy chief of staff for operations, Headquarters 5th Air Force, Yokota Air Base, Japan
14. June 1990 - July 1991, commander, 8th Tactical Fighter Wing, Kunsan Air Base, South Korea
15. August 1991 - May 1992, deputy director, operational requirements, Deputy Chief of Staff for Plans and Operations, Headquarters U.S. Air Force, the Pentagon, Washington, D.C.
16. June 1992 - February 1993, executive officer to the chief of staff, Headquarters U.S. Air Force, the Pentagon, Washington, D.C.
17. July 1993 - February 1995, commander, 49th Fighter Wing, Holloman Air Force Base, N.M.
18. February 1995 - May 1996, commander, 57th Wing, Nellis Air Force Base, Nev.
19. June 1996 - April 1998, chief of staff, Headquarters Allied Forces North Europe, NATO, Stavanger, Norway
20. April 1998 - May 1999, director for plans and policy, J-5, U.S. Atlantic Command, Norfolk, Va.
21. May 1999 - present, chief of staff, U.S. Atlantic Command, Norfolk, Va.

FLIGHT INFORMATION

Rating: Command pilot
Flight hours: More than 3,600
Aircraft flown: F-105, A-7D, F-16 and F-117

MAJOR AWARDS AND DECORATIONS

Defense Distinguished Service Medal

Legion of Merit with oak leaf cluster
Distinguished Flying Cross
Meritorious Service Medal with four oak leaf clusters
Air Medal with 11 oak leaf clusters
Vietnam Service Medal
Republic of Vietnam Gallantry Cross with palm
Republic of Vietnam Campaign Medal
Norwegian Distinguished Service Medal with Laurel

EFFECTIVE DATES OF PROMOTION
Second Lieutenant May 4, 1968
First Lieutenant Nov 4, 1969
Captain May 4, 1971
Major Sep 1, 1979
Lieutenant Colonel Feb 1, 1984
Colonel Jun 1, 1988
Brigadier General Jul 1, 1993
Major General Aug 4, 1996
(Current as of January 2000)

MAJOR GENERAL HARRY FALLS JR.

Retired April 1, 1985

Major General Harry Falls Jr. is deputy inspector general, Headquarters U.S. Air Force, Washington, D.C.

General Falls was born in Kings Mountain, N.C., and graduated from Lee H. Edwards High School, Asheville, N.C. He graduated from Clemson (S.C.) University in 1949 with a bachelor of science degree in agriculture and completed Air Command and Staff College at Maxwell Air Force Base, Ala., in 1961.

He entered aviation cadet training at Goodfellow Air Force Base, Texas, in June 1949 and graduated from Williams Air Force Base, Ariz., in June 1950 with his pilot wings and commission as a second lieutenant.

From June 1950 to September 1953, General Falls served as a fighter pilot and flight commander at Furstenfeldbruck Air Base, Germany. He then joined the Air Training Command, serving first at James Connally Air Force Base, Texas, as an instructor pilot and later at Laredo Air Force Base, Texas, as chief, standardization and evaluation, and squadron operations officer. From November 1957 to December 1958, he was flight commander and operations officer at Osan Air Base, South Korea. He returned to the United States and served with the 33rd Air Division at Richards-Gebaur Air Force Base, Mo., as a staff officer until September 1960.

After graduating from Air Command and Staff College, he was assigned as a flying safety officer at Headquarters Air Training Command, Randolph Air Force Base, Texas. From June 1964 to June 1967, he was F-5 project officer with the Military Assistance Advisory Group in Norway.

He completed the U.S. Air Force Operational Training Course at Davis-Monthan Air Force Base, Ariz., in April 1968. Subsequently he served a one-year tour of duty at Cam Ranh Bay Air Base, Republic of Vietnam, as an F-4 squadron operations officer and squadron commander. In the spring of 1969, he was assigned to Headquarters Pacific Air Forces, Hickam Air Force Base, Hawaii, as an operations staff officer in the Force Plans and Strategy Directorate, Office of the Deputy Chief of Staff, Plans.

The general transferred to Laughlin Air Force Base, Texas, in July 1972 as deputy commander of operations for the 47th Flying Training Wing and assumed command of the wing in June 1973. He was named assistant deputy chief of staff for operations, Headquarters Air Training Command, in July 1975. General Falls became deputy commander, 5th Allied Tactical Air Force, Vicenza, Italy, in August 1976. In May 1979 he became assistant for readiness, Headquarters U.S. Air Forces in Europe, Ramstein Air Base, Germany.

From August 1979 to January 1980, he commanded the 86th Tactical Fighter Wing, also at Ramstein Air Base, and the Kaiserslautern Military Community, the largest community of Army and Air Force personnel outside the continental United States. He then served as deputy chief of staff for operations and intelligence at U.S. Air Forces in Europe headquarters until March 1981.

General Falls returned to the United States and was assigned to Norton Air Force Base, Calif., as deputy inspector general for inspection and safety, and commander of the Air Force Inspection and Safety Center. He assumed his present duties in July 1982.

The general is a command pilot. His military decorations and awards include the Distinguished Service Medal, Legion of Merit with oak leaf cluster, Distinguished Flying Cross with oak leaf cluster, Bronze Star Medal, Air Medal with 15 oak leaf clusters, Joint Service Commendation Medal, and Air Force Outstanding Unit Award ribbon with "V" device and oak leaf cluster.

He was promoted to major general Feb. 4, 1980, with date of rank July 1, 1976.
(Current as of September 1982)

MAJOR GENERAL G. B. GREENE JR.

Retired Aug. 1, 1970

Major General G.B. (Ben) Greene Jr., is commander of Lackland Military Training Center, Air Training Command, and Lackland Air Force Base, San Antonio, Texas. In the dual capacity he is responsible for the operation of one of the largest installations in the Air Force as well as for the direction of a wide range of training functions at the base. The center provides basic training for new airmen and pre commissioning training for college graduates selected to enter officer career fields. In addition to the Basic Military School and the Officer Training School, the center also includes the U.S. Air Force Marksmanship School and the 3275th Technical School. The latter provides cryptographic, recruiting, security police, and other specialized training for officers and airmen.

General Greene was born in Laurens, S.C., in 1914. After graduating from high school in Anderson, S.C., he attended Clemson University, obtaining his bachelor of science degree in 1935.

He was appointed a second lieutenant in the Organized Reserve Corps, June 4, 1935, enlisted as an aviation cadet in January 1937, and completed flying training and received his pilot wings at Kelly Field, Texas, in January 1938.

General Greene was then assigned to the 8th Pursuit Group, Langley Field, Va., as a pilot and group engineering officer. He became commander of the 35th Pursuit Squadron in December 1941 and moved with it to the Pacific Theater of Operations, serving in New Guinea. He flew 68 combat missions totaling 200 combat hours, all in fighter aircraft, and destroyed three enemy aircraft in air-to-air combat, four on the ground, and one on the water.

General Greene was awarded the Distinguished Service Cross by General MacArthur for extraordinary heroism displayed over New Guinea on April 30, 1942. He received the Silver Star for gallantry in action over Port Morseby, New Guinea, and was awarded the Air Medal for meritorious achievement while participating in aerial flights in the Southwest Pacific area.

Upon his return to the United States in 1943,General Greene assumed command of the 423d Reconnaissance Group at De Ridder, La. In September 1943 he was transferred to Drew Field, Fla., as Assistant A-3 (Operations) for the 3d Fighter Command and later was assigned to the Third Air Force as deputy director of operations and training. In 1946 the Third Air Force became the Tactical Air Command with headquarters at Langley Field, Va., and he served as assistant chief of staff, A-3, of TAC until October 1947.

General Greene next was assigned to the 6th Fighter Wing in the Canal Zone as assistant chief of staff, A-3 (Operations). From August 1948 to August 1950 he served with the Caribbean Air Command. His next assignment took him to Stewart Air Force Base, N.Y., where he served with the Eastern Air Defense Forces.

In August 1951 he went to New Castle County Airport, Del., as commander of the 113th Fighter Interceptor Wing and later commanded the 4710th Defense Wing. In August 1952 he entered the Air War College at Maxwell Air Force Base, Ala., and after graduation in July 1953 he was assigned as commander of the 4708th Defense Wing at Selfridge Air Force Base, Mich.

He became deputy chief of staff for personnel, Air Defense Command, in July 1956; and he was named chief of staff of ADC in July 1959. Moving to Ramstein Air Base, Germany, in August 1960,he became deputy commander of the Seventeenth Air Force. In August 1961 he was named deputy chief of staff for operations at Headquarters, U.S. Air Forces in Europe, Wiesbaden, Germany.

General Greene returned to the United States and became director of military personnel, Deputy Chief of Staff for Personnel, at U.S. Air Force Headquarters in July 1963. He moved to the U.S. Air Force Military Personnel Center at Randolph Air Force Base, Texas, in July 1965,to become the assistant deputy chief of staff/personnel for military personnel, the new designation for the military functions which had been moved from Washington to Randolph.

In July 1967 he was assigned as commander of the Lackland Military Training Center, Air Training Command, and as commander of Lackland Air Force Base, Texas.

His decorations include the Distinguished Service Medal, Distinguished Service Cross, Silver Star, Legion of Merit and the Air Medal.

(Current as of Oct. 11, 1967)

MAJOR GENERAL CLIFTON D. WRIGHT

Retired March 1, 1986

Major General Clifton D. Wright Jr. is director of engineering and services, Office of the Deputy Chief of Staff, Logistics and Engineering, Headquarters U.S. Air Force, Washington, D.C.

General Wright was born in 1932, in New York City and grew up at various military installations. These included Langley Air Force Base, Va.; Fort McClellan, Ala.; Wiesbaden Air Base, Germany; and Eglin Air Force Base, Fla. Upon graduation from high school at Fort Walton Beach, Fla., the general attended Clemson (S.C.) University graduating in 1955 with a bachelor's degree in architectural engineering. He has a master's degree in management from The George Washington University, Washington, D.C., and is a graduate of the Air Command and Staff College, Maxwell Air Force Base, Ala., and the Industrial College of the Armed Forces, Fort Lesley J. McNair, Washington, D.C.

He entered the U.S. Air Force in April 1955 and was assigned to the 374th Installations Squadron, Tachikawa Air Base, Japan, as chief, Contract Inspection Section, and later as chief, Engineering and Construction Branch. General Wright's next assignment was to the Air Force Ballistic Missile Division, Camp Cooke (now Vandenberg Air Force Base),

Calif., in April 1957. There he was deputy base civil engineer during initial stages of base development. He later served as project officer for Capehart Housing construction and as Atlas weapon system project officer with Headquarters 1st Missile Division.

In October 1961 General Wright went to Italy for duty as base civil engineer at San Vito dei Normanni Air Station. Upon his return to the United States in April 1962, he was assigned as base civil engineer for Manzano Base, N.M., where he remained until he entered the Air Command and Staff College in 1966.

Following graduation in 1967, he was again assigned overseas, this time for a one-year tour of duty with Headquarters Military Assistance Command, Vietnam. In July 1968 he transferred to Headquarters Field Command (Defense Atomic Support Agency), Sandia Base, N.M., as command civil engineer in the Office of the Deputy Chief of Staff, Logistics. From 1971 to 1972, General Wright attended the Industrial College of the Armed Forces and concurrently earned a master's degree.

From August 1972 to February 1973, he was assigned to Headquarters U.S. Air Forces in Europe at Lindsey Air Station, Germany, as director of operations and maintenance. He then became assistant deputy chief of staff, engineering and services, and in April 1974 he was named the command's deputy chief of staff, engineering and services.

Returning to the United States in August 1975, he was assigned to Headquarters Strategic Air Command, Offutt Air Force Base, Neb., as deputy chief of staff, engineering and services. From August 1978 to January 1980 he served as the first commander of the Air Force Engineering and Services Center at Tyndall Air Force Base, Fla. He then moved to Air Force headquarters, Office of the Deputy Chief of Staff for Engineering and Logistics, as deputy director of engineering and services. General Wright assumed his present duties in August 1982.

He is a registered professional engineer in the state of New Mexico.

The general wears the senior missile badge. His military decorations and awards include the Legion of Merit with two oak leaf clusters, Bronze Star Medal, Meritorious Service Medal, Joint Service Commendation Medal, Air Force Commendation Medal with oak leaf cluster and Republic of Vietnam Armed Forces Honor Medal.

He was promoted to major general Jan. 1, 1982, with date of rank Aug. 1, 1978.
(Current as of August 1982)

BRIGADIER GENERAL ROBERT B. PLOWDEN JR.

Retired May 1, 1986

Brigadier General Robert B. Plowden Jr. is deputy director of operations, Headquarters U.S. Air Force, Washington, D.C.

General Plowden was born in 1932, in Sumter, S.C., where he attended public schools. He received a bachelor's degree from Clemson (S.C.) University in 1954 and a commission through the Reserve Officer Training Corps program.

After undergraduate pilot training at Spence Air Base, Ga., and Greenville Air Force Base, Miss., he earned his wings in September 1955. General Plowden completed combat crew training at Tyndall Air Force Base, Fla., and reported to his first operational assignment at George Air Force Base, Calif., in March 1956, flying F-86D's. He was assigned to Elmendorf Air Force Base, Alaska, from 1959 to 1963; and Castle Air Force Base, Calif., from 1963 to 1968, flying F-102s and F-106s.

Following RF-4C training at Shaw Air Force Base, S.C., he served a tour of duty in Southeast Asia with the 12th Tactical Reconnaissance Squadron, Tan Son Nhut Air Base, Republic of Vietnam, from April 1969 to April 1970, flying 440 combat hours. General Plowden returned to the United States as an Air Staff action officer and division chief in the Directorate of Plans at Air Force headquarters. He also served as an executive officer for the assistant to the secretary of the Air Force for international affairs and in June 1976 was assigned to the National Security Council staff at the White House.

In August 1977 General Plowden began the first of five assignments in Germany. Initially as commander of the 86th Combat Support Group, Ramstein Air Base and then as commander of the 86th Tactical Fighter Group in April 1978. From March 1979 to February 1980, he served as vice commander of the 36th Tactical Fighter Wing at Bitburg Air Base and then took command of the 52nd Tactical Fighter Wing, Spangdahlem Air Base. In February 1981 he was assigned as deputy chief of staff for operations, 2nd Allied Tactical Air Force, Rheindahlen Garrison. The general returned to Air Force headquarters in July 1983 and served as deputy director for regional plans and policy, Directorate of Plans. He assumed his present duties in September 1985.

A command pilot with 4,400 hours flying time, General Plowden's military decorations and awards include the Legion of Merit with two oak leaf clusters, Distinguished Flying Cross with two oak leaf clusters, Meritorious Service Medal, Air Medal with 11 oak leaf clusters and Air Force Commendation Medal.

He was promoted to brigadier general July 12, 1981, with same date of rank.

(Current as of September 1985)

BRIGADIER GENERAL LOUIS G GRIFFIN

Retired Feb. 1, 1970. Died Jan. 26, 1996

Brigadier General Louis G. Griffin is commander, 63rd Military Airlift Wing (Military Airlift Command), Norton Air Force Base, Calif.

General Griffin was born in Anderson, S.C., in 1918. He attended Anderson Boys High School and Clemson College, Clemson, S.C. He began his military career in 1939, when he entered the Army Air Corps as an aviation cadet. In November of that year, he graduated from flight training, was commissioned a second lieutenant and assigned to Maxwell Field, Ala., as an instructor.

He was assigned to Randolph Field, Texas, from April 1942 to January 1944 as a group commander and director of flying. For the next three months he served in the European Theater of Operations in England. From May 1944 until February 1945, he was group commander of the Single Engine Fighter Instructor School, Randolph Air Force Base, Texas. For the next year he was director of flying, Air Force Central Instructor School, Waco, Texas.

He next was assigned to Italy as executive officer and then as commander of the 62d Troop Carrier Group at Naples and later Pisa.

From January 1948 to July 1950, he served as commander of the 501st Maintenance and Supply Group, then the 3841st Maintenance and Supply Group and later the 3851st Air Base Group at Craig Air Force Base, Ala. During this period, he attended and graduated in 1949 from the Air Command and Staff College at Maxwell Air Force Base, Ala.

In July 1950 he was reassigned to Headquarters Third Air Force in England where he served as chief of Plans Development Branch and then deputy chief of staff, plans, where he remained until June 1953.

For the next three years General Griffin was assigned to Headquarters U.S. Air Force as chief of Programs Division and later deputy chief of staff, installations. He attended the Air War College during 1956-1957 and upon graduation remained at the Air University as secretary and chief of staff of the Air Command and Staff School for two years. From July 1959 to August 1960, he was deputy of Plans and Evaluation at the Air War College.

He then attended the Industrial College of the Armed Forces at Fort McNair, Washington, D.C., and upon graduation in August 1960 was assigned to the Military Air

Transport Service as vice commander of the 1502d Air Transport Wing, Hickam Air Force Base, Hawaii. He assumed command of the wing in April 1962.

In August 1964, he was assigned as chief of staff, Western Transport Air Force (MATS), Travis Air Force Base, Calif. General Griffin was assigned to Military Airlift Command headquarters in 1966 as deputy chief of staff, Personnel. He assumed his present duties as commander, 63rd Military Airlift Wing in August 1968.

He is a command pilot and his military decorations include the Legion of Merit with two oak leaf clusters and the commendation ribbon with oak leaf cluster.
(Current as of August 1, 1968)

BRIGADIER GENERAL CHARLES D. "RICK" ETHREDGE

Brig. Gen. Charles D. "Rick" Ethredge is Deputy to the Chief of Air Force Reserve, Headquarters U.S. Air Force, Washington, D.C. In this role, General Ethredge assists the Chief of Air Force Reserve who serves as the principal adviser on Reserve matters to the Air Force Chief of Staff.

General Ethredge received his commission in 1969 through the ROTC program at Clemson University, where he received a bachelor's degree in economics. He earned his wings the following year, first flying as a C-141 co-pilot with the 58th Military Airlift Squadron, and then as a UH-1N helicopter pilot in South Vietnam. The general later served as flight examiner, operations officer and UH-1H pilot before separating from active duty in July 1976. He continued his aviation career in the Reserve, including assignments as air reserve technician, C-141 flight examiner and C-130H pilot. His commands include the 315th Operations Group at Charleston AFB, S.C., the 913th Airlift Wing at Willow Grove Air Reserve Station, Pa., and the 403rd Wing at Keesler AFB, Miss.

General Ethredge is a command pilot with 7,600 flying hours in a variety of aircraft.

EDUCATION

1969 Bachelor's degree in economics, Clemson University, S.C.
1982 Squadron Officer School, by correspondence
1989 Air Command and Staff College, by correspondence
1993 Air War College, Maxwell Air Force Base, Ala.

ASSIGNMENTS

1. May 1969-May 1970, student, Undergraduate Pilot Training, Laredo AFB, Texas.
2. May 1970-August 1971, C-141 co-pilot, 58th Military Airlift Squadron, Robins AFB, Ga.
3. August 1971-April 1972, UH-IN helicopter pilot, 20th Special Operations Squadron, Cam Rahn Bay, South Vietnam
4. April 1972-July 1974, functional check and flight examiner, 317th Special Operations Squadron, Hurlburt Field, Fla.
5. July 1974-August 1975, rescue coordinator qualified in the T-39, Joint Coordination Center, Nakhon Phanom Royal Thai Air Base, Thailand
6. August 1975- July 1976, operations officer and UH-1H pilot, Detachment 1, 37th Aerospace Rescue and Recovery Squadron, Davis Monthan AFB, Ariz.
7. December 1978-February 1982, C-141 flight examiner pilot in lead airdrop and air refueling, 707th Military Airlift Squadron (associate), Charleston AFB, S.C.
8. February 1982-October 1993, air reserve technician, wing safety officer and aircraft operations officer, 315th Military Airlift Wing (associate), Charleston AFB, S.C.
9. October 1993-March 1994, air reserve technician, deputy operations group

commander and C-130H pilot, 94th Airlift Wing (associate), Dobbins Air Reserve Base, Ga.
10. March 1994-March 1998, air reserve technician, later, Commander, 315th Operations Group, Charleston AFB, S.C.
11. March 1998-April 2000, Commander, 913th Airlift Wing, Willow Grove Air Reserve Station, Pa.
12. April 2000-July 2004, Commander, 403rd Wing, Keesler AFB, Miss.
13. July 2004-present, Deputy to the Chief of Air Force Reserve, Headquarters U.S. Air Force, Washington, D.C.

FLIGHT INFORMATION
Rating: Command pilot
Flight hours: 7,600
Aircraft flown: C-141, UH-1N/H, HH-43, T-39, C-130H, C-17 and WC-130J

MAJOR AWARDS AND DECORATIONS
Legion of Merit
Meritorious Service Medal with oak leaf cluster
Joint Service Commendation Medal
Air Force Commendation Medal with oak leaf cluster
Joint Meritorious Unit Award
Air Force Outstanding Unit Award with silver and bronze oak leaf clusters
Combat Readiness Medal with two oak leaf clusters
National Defense Service Medal with bronze star
Armed Forces Expeditionary Medal with four bronze stars
Vietnam Service Medal
Southwest Asia Service Medal with two bronze stars
Humanitarian Service Medal with bronze star
Air Force Overseas Award Short
Air Force Longevity Service Award with silver and bronze oak leaf clusters
Armed Forces Reserve Medal
Small Arms Expert Marksmanship Ribbon
Air Force Training Ribbon
Republic of Vietnam Gallantry Cross with Palm
Republic of Vietnam Campaign Medal
Kuwait Liberation Medal (Kingdom of Saudi Arabia)
Kuwait Liberation Medal (Government of Kuwait)

EFFECTIVE DATES OF PROMOTION
Second Lieutenant May 9, 1969
First Lieutenant Nov. 9, 1970
Captain May 16, 1972
Major May 9, 1983
Lieutenant Colonel May 9, 1990
Colonel May 1, 1994
Brigadier General Jan. 30, 2003
(Current as of August 2004)

BRIGADIER GENERAL BASKIN R. LAWRENCE JR.

Retired Feb. 15, 1964. Died Feb. 2, 1980.

Baskin Reid Lawrence Jr., was born in 1911 in Seneca, S.C. He graduated from Seneca High School in 1928, and Clemson A&M College in 1932, receiving a bachelor's degree in Electrical Engineering and a second lieutenant commission in the Infantry Reserve. He entered the Army Air Corps Flying School in March 1934 where he successfully completed pilot training as a fighter pilot. His first assignment after Kelly Field was the 1st Pursuit Group at Selfridge Field, Mich.

In 1938 he was transferred to one of the Air Corps' first heavy reconnaissance squadrons at Langley Field, Va., which later moved to Miami, Fla. This squadron was later converted to a heavy bomb squadron assigned to the 91st Bomb Group which was sent to the 8th Air Force in England in September 1942. Colonel Lawrence later commanded this group, and completed 11 combat missions and 59 combat hours.

In late 1943 Colonel Lawrence was given the task of activating the first "Pathfinder Group," the 482nd, which was equipped with the first radar bomb system in the Air Corps, called the H2X. He received twelve H2X-equipped B-17s with specially trained crews who had no combat experience. When Colonel Lawrence requested time to combat-train these crews, the commander of the 8th Bomber Command answered, "You will train your crews over Germany, and the 8th Air Force will follow you dropping bombs when you do." For his performance in the 8th Air Force, he was awarded the Silver Star, Air Medal, and numerous citations including the Presidential Unit Citation.

At the end of World War II, Colonel Lawrence entered the Air Force Institute of Technology completing a two-year course in aeronautical engineering (electronics). He was then assigned to the Engineering Division, Air Materiel Command, as chief of the Materiels Laboratory, and later as deputy chief of plans and operations, Wright Air Development Center, Ohio.

He attended the Air War College in 1952-53, from which he was assigned to the Far East Air Forces as assistant deputy chief of staff, intelligence.

During his Far East tour, Colonel Lawrence was appointed to head a Joint Department of Defense/Department of State investigation into the interception of a U.S. Air Force RB-29 by a Russian MIG aircraft on Nov. 7, 1954. His report of this incident was accepted as the U.S. position in an extremely delicate international situation involving the governments of the United States, Japan and Russia. For this and his manner of performance at Far East Air Forces, he was awarded the Legion of Merit.

In January 1956 he returned to the United States and was assigned as deputy director, research and development, Headquarters U.S. Air Force, and in May 1960 became director of technology of the Air Force Command and Control Development Division at L.G. Hanscom Field, Bedford, Mass. General Lawrence remained as director of technology in the newly organized Electronic Systems Division of the Air Force Systems Command. He was promoted to the rank of brigadier general March 7, 1961.

On July 28, 1961, he assumed command of the U.S. Logistics Group, U.S. Air Forces in Europe, Ankara, Turkey.

Returning from overseas, General Lawrence was assigned duty as chief of the Air Defense Command's Command Control Defense Systems Office at L.G. Hanscom Field, Bedford, Mass., Aug. 16, 1963. In this capacity, he is responsible to the commander, ADC, through the chief of staff, for providing approved ADC guidance, advice and recommendations to the Electronic Systems Division and those military and civilian agencies under ESD supervision, and to keep the Commander, ADC, informed of all significant developments or studies on command and control systems with a possible aerospace defense application. (Up to date as of October 1963)

BRIGADIER GENERAL HENRY C. MORROW

General Hank Morrow is Assistant Adjutant General, J-3, JFHQ-KY, Kentucky National Guard, Frankfort, Kentucky. He also serves as the ANG Advisor to the USAF Assistant Surgeon General for Expeditionary Operations, Science and Technology and on the DOD Directed Energy Weapons Collection and Analysis Task Force for the Defense Intelligence Agency. General Morrow received his Air Force commission through the Reserve Officer Training Corps program at Clemson University, Clemson, South Carolina, in 1977.

He continued his education in Europe at Embry Riddle Aeronautical University, completing a master's degree in Aeronautical Science in 1984. An outstanding graduate of undergraduate pilot training, General Morrow has served as an F-111 and F-16 instructor pilot. His flying assignments include two tours in Europe, one in the F-111F, and one in the F-16.

After completing the USAF F-16 Fighter Weapons School in 1985, he was assigned to Hill AFB, Utah, and served as both a squadron and wing weapons officer. His previous experience with the Texas Air National Guard includes assignments as 182th Fighter Squadron Commander, 149FW Operations Group Commander, Director of Operations, HQ TXANG, Vice Commander, 147th Fighter Wing, Houston, Texas and Wing Commander, 149th Fighter Wing, San Antonio, Texas.

He joined the KYANG in March 2002 and previously served as the Chief of Staff, HQ KYANG. General Morrow is a combat veteran and command pilot with more than 3,000 flying hours in trainer and fighter aircraft.

EDUCATION:

1977 Bachelor of Science degree in Administrative Management, Clemson University, SC

1978 USAF Pilot Training, Vance AFB, OK

1984 Squadron Officer School, by correspondence

1984 Masters of Aeronautical Science, Embry Riddle Aeronautical University 1985 U.S. Air Force Fighter Weapons Instructor School, Nellis Air Force Base, NV

1993 Air Command and Staff College, by correspondence

1997 Air War College, by correspondence

ASSIGNMENTS:

1. April 1978 - May 1979, student, undergraduate pilot training, Vance Air Force Base, OK
2. May 1979 - July 1979, student, fighter lead in training, Holloman Air Force Base, NM
3. July 1979 - December 1979, student, F111D upgrade training, Cannon Air Force Base, NM
4. December 1979 - December 1981, Instructor Pilot F-111F, RAF Lakenheath, United Kingdom
5. December 1981 - May 1982, student pilot, F-16A upgrade training,MacDill Air Force Base, FL
6. May 1982 - December 1984, Instructor Pilot Flight Examiner F-16A, 50th TFW, Wing Executive Officer, Hahn Air Base, Germany
7. December 1984 - May 1985, student, U.S. Air Force Weapons Instructor Course, Nellis Air Force Base, NV
8. May 1985 - May 1988, Instructor Pilot, Chief of Weapons, 4th TFS and 16th TFS, Assistant Chief of Weapons, 388th TFW, Hill Air Force Base,

UT
9. July 1988 - December 1989, Instructor Pilot Weapons and Tactics officer, 182 th TFS (TXANG), Kelly Air Force Base, TX
10. December 1989 - July 1991, Instructor Pilot Weapons and Tactics officer, 111th FIS (TXANG), Ellington Field, Houston, TX
11. July 1991 - November 1991, Instructor Pilot, Chief of Quality Assurance, 147 CAMS, Ellington Field, Houston, TX
12. November 1991 - November 1992, Instructor Pilot and Weapons and Tactics officer, 182th TFS (TXANG), Kelly Air Force Base, TX
13. November 1992 - July 1994, Assistant Flight Commander, Instructor Pilot and Weapons and Tactics Officer, 182th FS, (TXANG), Kelly Air Force Base, TX
14. July 1994 - November 1995, Chief of Standardization and Evaluation, 149th Fighter Wing, Kelly Air Force Base, TX
15. November 1995 - April 1997, Commander, 182d Fighter Squadron, Kelly Air Force Base, TX
16. April 1997 - October 1998, Commander, 149th Operations Group, Kelly Air Force Base, TX
17. October 1998 - June 1999, Director Of Operations, HQ Texas Air National Guard, Camp Mabry, Austin, TX
18. June 1999 - February 2000, Vice Commander, 147th Fighter Wing, Ellington Field, TX
19. February 2000 - March 2002, Commander, 149th Fighter Wing, Lackland AFB, TX
20. March 2002 - Aug 2003, Chief of Staff, HQ Kentucky Air National Guard, Frankfort, KY
21. Aug 2003 - Nov 2003, Director, Combined Air Operations Center (CAOC), CENTAF, Al Udeid AB, Qatar

FLIGHT INFORMATION:
Rating: Command Pilot
Flight hours: more than 3,000 military hours and 7500 civilian hours
Aircraft flown: T-37, T-38, F-111, and F-16

MAJOR AWARDS AND DECORATIONS:
Legion of Merit
Air Force Meritorious Service Medal
Air Force Air Medal
Air Force Commendation Medal with one oak leaf cluster
Air Force Achievement Medal with two oak leaf clusters
Air Force Outstanding Unit Award with two oak leaf clusters
Air Force Organizational Excellence Award
Combat Readiness Medal with one silver and three bronze oak leaf clusters
Air Force Overseas Ribbon - Long with one oak leaf cluster
Air Force Longevity Service Award Ribbon with four oak leaf clusters
Armed Forces Reserve Medal with mobilization device
Texas Lone Star Distinguished Service Medal with two devices
Texas Medal of Merit with two devices
Texas Adjutant Generals Individual Award with one device
Texas Faithful Service Medal with one device

EFFECTIVE DATES OF PROMOTION:

Second Lieutenant, 22 December 1977
First Lieutenant, 14 May 1980
Captain, 14 May 1982
Major, 26 September 1991
Lieutenant Colonel, 15 October 1995
Colonel, 30 July 1999
Brigadier General, 23 June 2003.

BRIGADIER GENERAL FISK OUTWATER

Retired effective Jan 9, 2000

Brigadier General Fisk Outwater is the assistant adjutant general who also serves as commander of the North Carolina Air National Guard. General Outwater is the primary advisor to the adjutant general of North Carolina on Air National Guard affairs both State and Federal. He is responsible for formulating plans for flying activities based on training requirements and directives from higher headquarters. He analyzes and evaluates training programs and operational readiness status to determine training effectiveness, mission capability, unit effectiveness, mission accomplishments, and problems pertaining to operations related matters. The North Carolina Air National Guard is comprised of units located at the 145th Airlift Wing, Charlotte/Douglas International Airport; 118th Combat Communications Squadron, Albemarle-Stanly County Airport; and 263rd Combat Communications Squadron, Badin, N.C.

The general began his career by receiving a commission through reserve officer training school in 1965. His first assignment was flying C-130 aircraft at Ching Chuan Kang Air Base, Taiwan, where he flew more than 100 combat missions and 250 hours combat time in Vietnam during 1967-1968. He is a command pilot with more than 5,000 flying hours.

EDUCATION:

1965 Bachelor of science degree in ceramic engineering, Clemson University, Clemson, S.C.
1972 Master's degree in ceramic engineering, Clemson University, Clemson, S.C.
1975 Squadron Officer School, by correspondence
1980 Air Command and Staff College, by correspondence
1986 Air War College, by correspondence

ASSIGNMENTS:

1. March 1967 - March 1968, student, undergraduate pilot training, Craig Air Force Base, Ala.
2. September 1968 - October 1969, 50th Tactical Airlift Squadron, Ching Chaun Kang, Republic of Taiwan
3. October 1969 - December 1970, 61st Tactical Airlift Squadron, Sewart Air Force Base, Tenn.
4. December 1970 - March 1971, 61st Tactical Airlift Squadron, Little Rock Air Force Base, Ark.
5. July 1972 - January 1979, pilot, 156th Tactical Airlift Squadron, Charlotte, N.C.
6. January 1979 - January 1980, assistant flight commander, 156th Tactical Airlift Squadron, Charlotte, N.C.
7. January 1980 - November 1987, flight commander C-130B standard/evaluation pilot, 156th Tactical Airlift Squadron, Charlotte, N.C.
8. December 1987 - October 1988, chief standard/evaluation tactical airlift C-130B,

145th Tactical Airlift Group, Charlotte, N.C.
9. October 1988 - May 1991, deputy commander of maintenance, Headquarters, 145th Tactical Airlift Group, Charlotte, N.C.
10. June 1991 - March 1992, air operations staff officer, Headquarters, 145th Tactical Airlift Group, Charlotte, N.C.
11. March 1992 - August 1995, commander, 145th Airlift Group, Charlotte, N.C.
12. August 1995 - July 1998, chief of staff who also serves as commander, Headquarters North Carolina Air National Guard, Charlotte, N.C.
13. July 1998 - January 2000, assistant adjutant general who also serves as commander, Headquarters North Carolina Air National Guard, Charlotte, N.C.

FLIGHT INFORMATION:

Rating: Command pilot
Flight hours: More than 5,454
Aircraft flown: C-130B, E, H-3, T-37, T-38
Pilot wings from: Craig Air Force Base, Ala.

MAJOR AWARDS & DECORATIONS:

Legion of Merit
Meritorious Service Medal
Air Medal with two oak leaf clusters
Air Force Commendation Medal
Air Force Achievement Medal
Air Force Outstanding Unit Award with four oak leaf clusters
Combat Readiness Medal with service star
National Defense Service Medal with service star
Armed Forces Expeditionary Medal
Vietnam Service Medal
Southwest Asia Service Medal with bronze star
Humanitarian Service Medal with bronze star
Air Force Longevity Service Award Ribbon with one silver and one bronze oak leaf cluster
Armed Forces Reserve Medal with hourglass device
Small Arms Expert Marksmanship Ribbon
Air Force Training Ribbon
Vietnam Campaign Medal
North Carolina National Guard Commendation Medal
North Carolina National Guard Meritorious Unit Citation
North Carolina National Guard State Active Duty Medal
North Carolina National Guard State Service Award with gold propeller

OTHER ACHIEVEMENTS:

Deputy director of mobility forces in support of Operation Restore Hope in Somalia, 1993
Mission commander for air and assigned to Reserve Component Forces for Operation Provide Promise in Bosnia, 1993
C-130 representative for the USAF/CC to the Joint Task Force for Provide Promise in Bosnia, 1993
IG Brown Command Excellence Award, 1994

CIVILIAN OCCUPATION:

Commercial airline pilot

EFFECTIVE DATES OF PROMOTION:

Second Lieutenant Dec 18, 1965
First Lieutenant Sep 20, 1967
Captain Mar 20, 1969
Major Jun 15, 1978
Lieutenant Colonel Feb 13, 1984
Colonel Jun 02, 1991
Brigadier General Sep 25, 1996

(Current as of February 2001)

BRIGADIER GENERAL CHALMERS R. CARR JR.

Retired Feb. 1, 1991

Brigadier General Chalmers R. Carr Jr. is inspector general, Headquarters Air Training Command, Randolph Air Force Base, Texas.

General Carr was born in Philadelphia in 1937, and graduated from Christ School, Arden, N.C., in 1956. He received a bachelor's degree in industrial management from Clemson University in 1960 and a master's degree in personnel management from Central Michigan University in 1976. He completed the Industrial College of the Armed Forces in 1973, Air Command and Staff College in 1976, and the Australian Joint Service Staff College in 1980.

The general was commissioned through the Reserve Officer Training Corps program in 1960. He completed undergraduate pilot training at Williams Air Force Base, Ariz., in 1962, and remained there as a T-33 and T-37 flight instructor, and as an officer training instructor.

In 1966 General Carr was assigned to the Military Advisory Group, Thailand, as operations adviser to the Royal Thai Air Force Flying School, Korat Royal Thai Air Force Base. In early 1968 he served as an exchange officer with the U.S. Navy's Flying Training Squadron Seven at Meridian Naval Air Station, Miss.

He transferred to the 432nd Tactical Reconnaissance Wing, Udorn Royal Thai Air Force Base, Thailand, in March 1971, flying F-4D Phantoms with the 555th Tactical Fighter Squadron. After completing a tour of duty as T-38 section commander at Columbus Air Force Base, Miss., in 1974, General Carr was assigned to the Current Operations and Briefing Branch, Director of Operations, Deputy Chief of Staff, Operations, Plans and Readiness, Headquarters U.S. Air Force, Washington, D.C. He was assistant executive officer to the deputy chief of staff, operations, plans and readiness, from July 1977 until March 1979, when he became executive officer.

In July 1980 General Carr was assigned to Alaska, where he was commander of both King Salmon Air Base and the 5071st Air Base Squadron. He reported to Laughlin Air Force Base, Texas, in July 1981, and served initially as deputy commander for maintenance for the 47th Flying Training Wing before assuming command of the 47th Air Base Group in November 1982. In July 1984 he assumed command of the 3480th Technical Training Wing, Goodfellow Air Force Base, Texas. Subsequently, General Carr became the first commander of the newly established Goodfellow Technical Training Center in March 1985. In April 1986 he became director of inspection, Air Force Inspection and Safety Center, Norton Air Force Base, Calif. He assumed his present position in February 1990.

The general is a command pilot with more than 4,020 flying hours.

His military awards and decorations include the Legion of Merit,

Distinguished Flying Cross with oak leaf cluster, Bronze Star Medal,

Meritorious Service Medal with two oak leaf clusters, Air Medal with nine oak leaf clusters, Air Force Commendation Medal, Navy Commendation Medal, Air Force Outstanding Unit Award with "V" device and five oak leaf clusters,
National Defense Service Medal,
Vietnam Service Medal with seven service stars,
Air Force Longevity Service Award Ribbon with five oak leaf clusters.

He was promoted to brigadier general April 1, 1986, with date of rank Oct. 1, 1985. (Current as of April 1990)

BRIGADIER GENERAL JOHN A. DES PORTES

Retired March 1, 1972. Died Jan. 14, 2001.

Brigadier General John A. Des Portes is commander of the 12th Strategic Aerospace Division, with headquarters at Davis-Monthan Air Force Base, Ariz. This division is the largest in the Strategic Air Command and the most diversified, having both intercontinental ballistic missiles and reconnaissance units.

General Des Portes was born in 1920 in Ridgeway, S.C., where he graduated from Ridgeway High School. He graduated from Clemson University, Clemson, S.C., with a bachelor of science degree and, as a member of the Reserve Officers Training Corps, his commission as a second lieutenant, Infantry, Army Reserve, in 1941. He entered active military service in August 1941 in the Army Air Corps, attended aviation cadet training, and in March 1942 graduated from Advanced Flying School at Brooks Field, Texas, with a commission as a second lieutenant in the U.S. Army Air Corps and his pilot wings.

In March 1942 General Des Portes was assigned as commander and adjutant for a training squadron at Tyndall Field, Fla.; in February 1943 was assigned as training group commander and director of flying; and in January 1944 became director of B-24 copilot training.

He was transferred to the European Theater of Operations in March 1945, served as commander of the 344th Bombardment Squadron, 98th Bombardment Group, in Italy, and flew 14 combat missions in B-24 bomber aircraft. At the end of World War II in Europe, he was assigned as squadron commander and operations officer at McCook AFB, NE and then Fort Worth, Texas.

In August 1946 he became operations officer for the Strategic Air Command Air Demonstration Unit at Smokey Hill Army Air Field, Kansas. He returned to Fort Worth, Texas, in December 1946, and served as Operations officer for the 436th Bombardment Squadron, 7th Bombardment Group. In July 1948 he became chief of the Bombardment Unit, Flying Training Branch, Directorate of Operations, at Headquarters, Strategic Air Command, then located at Andrews Air Force Base, Washington, DC. He stayed with the command when it moved to Offett Air Force Base in November 1948. In July 1950 he entered the Air Command and Staff School at Maxwell Air Force Base, Ala.

In December 1950 General Des Portes was assigned as operation officer with the 91st Strategic Reconnaissance Group, Barksdale Air Force Base, La. This group was equipped with the RB-45, Strategic Air Command's first multiengine jet and moved to Lockbourne Air Force Base, Ohio, in 1951 where he served as squadron commander, wing operations officer, and deputy commander for the 91st Strategic Reconnaissance Wing. In May 1954 he was named director of plans, Second Air Forcc, at Barksdale Air Force Base.

General Des Portes was assigned to Headquarters U.S. Air Force in the Pentagon in June 1957 in the Office of the Deputy Chief of Staff, Operations, and became chief of the Strategic Division, Directorate of Operations. He entered the National War College,

Washington, D.C., in August 1960, and in July 1961 was transferred to Laughlin Air Force Base, Texas, as commander of the 4080th Strategic Wing, the only Air Force organization equipped with the U-2 aircraft. The wing was the first to bring back conclusive proof of offensive weapons in Cuba during the October 1962 Cuban Missile Crisis.

President John F. Kennedy personally presented the 4080th Wing with the first oak leaf cluster to the Air Force Outstanding Unit Award in November 1962. The wing moved to Davis-Monthan Air Force Base, Arizona in July 1963, and General Des Portes remained in command until June 1965, when he was transferred to Beale Air Force Base, California as the commander of the 4200th Strategic Reconnaissance Wing. This wing is the only Air Force organization with the SR-71 reconnaissance aircraft. He assumed command of the 14th Strategic Aerospace Division at Beale Air Force Base, in September 1965, and became commander of the 47th Air Division at Castle Air Force Base, CA, in December 1966. General Des Portes served from July 1968 until September, 1970 as chief of staff, U.S. Taiwan Defense command, at Taipei, Taiwan. and assumed duties as commander of the 12th Strategic Aerospace Division with headquarters at Davis-Monthan Air Force Base, Arizona in October, 1970.

His military decorations include the Legion of Merit with two oak leaf clusters, Air Medal, Air Force Commendation Medal with oak leaf cluster, Army Commendation Medal, Presidential Unit Citation emblem, and Air Force Outstanding Unit Award Ribbon with two oak leaf clusters.

He is a Command Pilot and his hometown is Fort Mill, S.C

He was promoted to the temporary grade of brigadier general effective Sept. 30, 1965, with date of rank Sept. 30, 1965.
(Current as of Dec. 15, 1970)

BRIGADIER GENERAL CHARLES H. KING
Retired 01 March 2001

Brigadier General Charles H. King was vice commander, Aeronautical Systems Center (ASC), Air Force Materiel Command, Wright-Patterson Air Force Base, Ohio.

General King was commissioned as a distinguished graduate of the Clemson University Air Force ROTC program. He entered active duty in 1967 after completing a master's degree program and served a four-year tour of duty as a civil engineering officer at various locations in the United States and the Pacific. During this time, he commanded a civil engineering squadron with the 355th Tactical Fighter Wing at Takhli Royal Thai Air Force Base, Thailand.

The general joined the Air Force Reserve upon separation from active duty in 1971. He was assigned as an individual mobilization augmentee (IMA) to the Aeronautical Systems Division at Wright-Patterson Air Force Base where he was an acquisition and civil engineering staff officer, and development engineering manager. He performed duty in several major system program offices, including the B-1B, B-52, F-16, and LANTIRN. He commanded an Aeronautical Systems Division Reserve Detachment at Arnold Air Force Base, Tennessee and later served as senior IMA for engineering in the Directorate of Science and Technology, Air Force Materiel Command.

Prior to his present assignment, General King served as mobilization assistant to the director, Science and Technology, Air Force Materiel Command, and as mobilization assistant to the Air Force Technology executive officer and commander, Air Force Research Laboratory.

EDUCATION:

1966 Bachelor of science degree in Civil Engineering, Clemson University, Clemson, S.C.
1967 Master of science degree in Civil Engineering, Clemson University, Clemson, S.C.
1974 Graduate study in management, East Tennessee State University
1978 Squadron Officer School, by correspondence
1981 Executive Development Program, Spring Hill College, Mobile, Ala.
1982 Air Command and Staff College, by correspondence
1985 Air War College, by seminar
1994 Program Managers Course, Defense Systems Management College, Fort Belvoir, Va.

ASSIGNMENTS:

1. October 1967 - September 1970, aerospace facilities engineer, Arnold Engineering Development Center, Arnold Air Force Base, Tenn.
2. September 1970 - December 1970, civil engineering staff officer, 355th Civil Engineering Squadron, Takhli Royal Thai Air Force Base, Thailand
3. December 1970 - March 1971, civil engineering staff officer, 6235th Air Base Squadron, Takhli Royal Thai Air Force Base, Thailand
4. March 1971 - April 1971, civil engineering staff officer, Headquarters 7th Air Force, Tan Son Nhut Air Base, South Vietnam
5. April 1971 - October 1971, construction management engineer, 354th Civil Engineering Squadron, Myrtle Beach Air Force Base, S.C.
6. October 1971 - April 1973, civil engineer officer, Headquarters Air Reserve Personnel Center, Denver, with duty at 9940th Air Reserve Squadron, Charlotte, N.C., and 9945th Air Reserve Squadron, Raleigh, N.C.
7. May 1973 - November 1973, procurement officer, Aeronautical Systems Division, Wright-Patterson Air Force Base, Ohio
8. November 1973 - November 1979, acquisition program management officer, Aeronautical Systems Division, Wright-Patterson Air Force Base, Ohio
9. November 1979 - April 1985, acquisition project officer, Aeronautical Systems Division, Wright-Patterson Air Force Base, Ohio
10. April 1985 - March 1990, Civil Engineering staff officer, Aeronautical Systems Division, Wright-Patterson Air Force Base, Ohio
11. March 1990 - August 1993, base development engineer structures, Aeronautical Systems Center, Wright-Patterson Air Force Base, Ohio
12. August 1993 - March 1995, senior IMA for engineering, directorate of science and technology, Headquarters Air Force Material Command, Wright-Patterson Air Force Base, Ohio
13. March 1995 - October 1997, mobilization assistant to the director, science and technology, Headquarters Air Force Materiel Command, Wright-Patterson Air Force Base, Ohio
14. November 1997 - April 2000, mobilization assistant to the commander, Air Force Research Laboratory, Wright-Patterson Air Force Base, Ohio
15. April 2000 - September 2000, mobilization assistant to the commander, Aeronautical Systems Center (ASC), Wright-Patterson Air Force Base, Ohio
16. October 2000 - present, vice commander, Aeronautical Systems Center (ASC), Wright-Patterson Air Force Base, Ohio

SPECIAL QUALIFICATIONS:

Master Acquisition Badge
Senior Civil Engineering Badge
Registered Professional Engineer
Certified DoD Acquisition Professional, Level III in Program Management,
Test and Evaluation, and Systems Planning, Research, Development and Engineering

MAJOR AWARDS AND DECORATIONS:

Legion of Merit
Meritorious Service Medal with oak leaf cluster
Air Force Commendation Medal
Air Force Achievement Medal
Air Force Outstanding Unit Award with "V" device
Vietnam Service Medal with two service stars
Armed Forces Reserve Medal with two devices
National Defense Service Medal with device
Republic of Vietnam Gallantry Cross with Palm

EFFECTIVE DATES OF PROMOTION:

Second Lieutenant May 14, 1966
First Lieutenant Oct 15, 1968
Captain Apr 15, 1970
Major May 14, 1980
Lieutenant Colonel September 30, 1985
Colonel Jul 1, 1989
Brigadier General Feb 29, 1996

(Current as of October 2000)

BRIGADIER GENERAL MICHAEL G. LEE

Retired Feb. 1, 2004.

Brig. Gen. Michael G. Lee is Military Executive, and Director for Military Support and Operations, National Imagery and Mapping Agency, Bethesda, Md. NIMA is a combat support agency jointly chartered by the secretary of defense and the director of central intelligence. The agency is the central authority for imagery and mapping in the U.S. government. It provides geospatial intelligence support to U.S. and allied military forces, other government activities and the civil sector in support of national security objectives.

General Lee entered the Air Force in 1973 as a graduate of the Clemson University Air Force ROTC program. He has served in various flying assignments involving the F-15, AT-38, T-38, T-37 and T-1. He has served on the Office of the Secretary of Defense, joint and air staffs. General Lee has commanded an operations group, the 14th Flying Training Wing, Columbus Air Force Base, Miss., and served as the Air Education and Training Command's Inspector General, Randolph Air Force Base, Texas.

EDUCATION

1972 Bachelor of arts degree in chemistry, Clemson University, Clemson, S.C.
1975 Squadron Officer School, by correspondence
1980 Master of arts degree in management, Troy State University, European Division
1987 Air Command and Staff College, Maxwell Air Force Base, Ala.
1988 Air War College, by correspondence
1992 Royal College of Defence Studies, London, England
1998 National Security Management Course, Syracuse University, Syracuse, N.Y.

ASSIGNMENTS

1. February 1973 - January 1974, student, undergraduate pilot training, Moody Air Force Base, Ga.
2. July 1974 - December 1975, T-38 instructor pilot, 71st Flying Training Squadron, Moody Air Force Base, Ga.
3. December 1975 - November 1977, T-38 instructor pilot, 50th Flying Training Squadron, Columbus Air Force Base, Miss.
4. November 1977 - September 1978, student, F-15 combat crew training, F-15 Replacement Training Unit, Luke Air Force Base, Ariz.
5. September 1978 - August 1981, F-15 pilot, 525th Tactical Fighter Squadron, Bitburg Air Base, West Germany
6. August 1981 - September 1983, F-15 instructor pilot, 60th Tactical Fighter Squadron, later, F-15 standardization and evaluation instructor pilot, 33rd Tactical Fighter Wing, Eglin Air Force Base, Fla.
7. September 1983 - May 1986, inspection planner, later, Chief, F-15 Inspection Branch, Inspector General Team, Headquarters Tactical Air Command, Langley Air Force Base, Va.
8. May 1986 - June 1987, student, Air Command and Staff College, Maxwell Air Force Base, Ala.
9. July 1987 - June 1989, U.S. Air Force representative to the Tactical Missile Defense Program and Chief, North Atlantic Treaty Organization Air Defense Section, later, Deputy Chief, Requirements Division, Directorate of Plans, Deputy Chief of Staff for Plans and Operations, Headquarters U.S. Air Force, Washington, D.C.
10. June 1989 - January 1992, U.S. Air Force Military Assistant to the Executive Secretary, later, Deputy Executive Secretary to the Secretary of Defense, Washington, D.C.
11. January 1992 - December 1992, student and instructor, Royal College of Defense Studies, London, England
12. January 1993 - August 1994, Commander, 47th Operations Group, Laughlin Air Force Base, Texas
13. August 1994 - August 1995, Inspector General, Headquarters Air Education and Training Command, Randolph Air Force Base, Texas
14. August 1995 - July 1997, Commander, 14th Flying Training Wing, Columbus Air Force Base, Miss.
15. July 1997 - August 1998, Executive Officer to the Deputy Commander in Chief, U.S. European Command, Stuttgart-Vaihingen, Germany
16. August 1998 - April 1999, Director of Inspections, Office of the Inspector General, Office of the Secretary of the Air Force, Washington, D.C.
17. April 1999 - March 2000, Deputy Director for Operations, National Military Command Center, the Joint Staff, Washington, D.C.
18. April 2000 - June 2001, Deputy Commander, Combined Air Operations Center 6, Allied Air Forces South, NATO, Eskisehir, Turkey
9. June 2001 - September 2001, Deputy Director, Plans and Customer Operations Directorate, National Imagery and Mapping Agency, Reston, Va.
20. September 2001 - May 2002, Executive Director, and Director for Military Support and Operations, National Imagery and Mapping Agency, Bethesda, Md.
21. May 2002 - present, Military Executive, and Director for Military Support and Operations, National imagery and Mapping Agency, Bethesda, Md.

FLIGHT INFORMATION

Rating: Command pilot
Flight hours: More than 2,800
Aircraft flown: F-15A/B/C/D, T-37, T-38A, T-1 and AT-38

MAJOR AWARDS AND DECORATIONS

Defense Superior Service Medal with three oak leaf clusters
Legion of Merit with oak leaf cluster
Bronze Star Medal
Defense Meritorious Service Medal with oak leaf cluster
Meritorious Service Medal with three oak leaf clusters
Air Force Commendation Medal
Air Force Achievement Medal
National Intelligence Medal of Achievement

EFFECTIVE DATES OF PROMOTION

Second Lieutenant Feb 6, 1973
First Lieutenant Feb 6, 1975
Captain Feb 6, 1977
Major Aug 1, 1984
Lieutenant Colonel Jul 1, 1988
Colonel Jan 1, 1992
Brigadier General Jan 1, 2000
(Current as of February 2004).

COLONEL WILLIAM L. LEVERETTE
WWII Fighter Ace

William L. Leverette was born in Palatka, Florida on September 5, 1913, and was graduated with a degree in Mechanical Engineering from Clemson A and M College and a Masters degree in Aeronautical Engineering from Princeton. He entered the Army Air Corps in 1939 and earned his pilots wings in 1940. As commander of the 37th Fighter Squadron, Leverette on October 9, 1943, led seven P-38s on a mission to protect the Royal Navy warships in the Mediterranean. When he sighted a formation of 30 German JU-87 Stuka dive bombers, Leverette sent three of his fighters to fly top cover while he and the three closed in on the Stukas. Each of the German aircraft had a gunner manning a flexible machine gun plus two wing mounted guns.

Leverette, who had spent two years teaching fighter tactics and had more that 1,000 hours flying fighters, took out seven of the Stukas himself. His unit shot down another nine plus a Ju-88, and most of the rest headed home or ended up in the sea.

For his leadership and individual performance in this action. he recieved the Distuinguished Service Cross.

He continued to shoot down enemy aircraft in other actions, and his final total was eleven victories(7 Stukas, two Me-109s, 2 Me 110s), ending his war as one of the 20 aces in the Mediterranean theater.

During his military career Col Leverette flew 45 different aircraft from the BT-2 biplane to the F-104. He retired from the USAF in 1965.

He died April 7, 2003 in Beaverton, Oregon and was interred in Arlington National Cemetary, Washington, DC.

WILLIAM H. JONES

Col USAFR-Ret and Finance Executive

Bill entered Clemson in 1951, and while a student majoring in Business Management, he decided to join the Flying Club, later becoming its president in 1953. He received his degree in Textile Manufacturing in 1955 and also his commission as a 2nd Lt in the USAFR thru the Air Force ROTC program. While waiting to be called to active duty he enrolled for Graduate Work at George Washington University but was recalled to attend USAF Pilot Training Class 56P before he could complete his studies. He entered training at Greenville AFB, Mississippi, completed his advance training, and received his pilots wings. He was returned to Greenville AFB where he instructed in the Lockheed T-33 aircraft. He next was transferred to James Conneley AFB, where he received further training, and then later to Randolph AFB, where he then became an Instrument Flying Instructor flying the Northrop T-38 and the North American T-39 Sabreliner. After a short time at Randolph he was assigned to the 460th Tactical Fighter Wing in Vietnam and began flying combat missions in the Phantom F-4 aircraft, completing 172 combat missions. He was transferred to the Atomic Energy Commission and was involved in the Atomic Energy Commission, involved in upper air nuclear test missions. He was transferred after his combat tour to a job supporting the Joint Chief of Staff in Washington, DC. Next he was again transferred and served in the Diplomatic Corp-American Embassy in Brussels, Belgium for 4 years as an aid to the U.S. Representative, North Atlantic Treaty Organization. He retired from active service and as a Col and then joined Merrill Lynch. He is presently a Vice President with Merrill Lynch. A member of the President's and Chairman's Club, a Certified Financial Planner-Graduate of Don Regan School of Advanced Financial Planning-Princton University.

Professional Associations

Rotary, Masons, Scottish Rite, Shrine, Order of Daedaleions,
and the Air Force Association
Served as Investment Advisor to University Club Advisory Board
Served as Board Member for Salvation Army
Served as Board Member for National Kidney Foundation
Served as Board Member for the Mental Health Association
President Elect Southwest Rotary Board
Listed in "Who's Who in America" in Finance and Industry
Listed in "Who's Who in the World" in Finance and Industry.

Military Decorations

Legion of Merit
Vietnam Service Medal
Republic of Vietnam Campaign Medal
Presidential Unit Citation
Air Medal with 5 Oak Leaf Clusters

He still is an active General Aviation pilot, having more than 2000 hours logged, and flies his own Cessna 340A aircraft for business and pleasure.

COLONEL CHRISTOPHER B. AYRES

Colonel Christopher B. Ayres is the 460th Operations Group Commander, 460th Space Wing, Buckley Air Force Base. The wing provides combatant commanders with superior global surveillance, worldwide missile warning, expeditionary forces and support to homeland defense missions. As base host, the wing provides support services to 38 associate organizations and approximately 88,000 people in the Front Range.

Colonel Ayres is a graduate of Clemson University and was commissioned through AFROTC in 1985 as a distinguished graduate. He has held key leadership positions in Ground Launch Cruise Missile (GLCM), Space, and Intercontinental Ballistic Missile (ICBM) squadrons, wings and theater headquarters staffs. His leadership positions have included GLCM crew commander and senior missile crew; space and missile warning Deputy Crew Commander, Crew Commander, and Flight Commander, ICBM Operations Officer, and ICBM Squadron Commander. Colonel Ayres has also had multiple assignments in joint operations while assigned to Headquarters, United States Space Command, Headquarters, United States European Command, and the Joint Staff.

Colonel Ayres is a command space and missile operator with more than 21 years of experience.

EDUCATION:

1985 Bachelor of Science, Engineering Technology, Clemson University.
1990 Squadron Officer's School, Maxwell Air Force Base, Ala.
1995 Masters of Science, Space Operations, University of Colorado.
1996 U.S Air Force Space Tactics School (AFSPC), Falcon Air Force Base, Colo.
2000 Masters of Science, Military Operational Art and Science, USAF Air Command and Staff College, Montgomery, Ala.
2004 Masters of Science, National Security Strategy, National War College, Fort Lesley J. McNair, Washington, D.C.

ASSIGNMENTS:

1. November 1985 - February 1996, Student, Ground Launch Cruise Missile, 868th Tactical Missile Training Squadron, Davis-Monthan AFB, Ariz.
2. February 1996 - January 1988, Ground Launch Cruise Missile Launch Control Officer, Senior Instructor Deputy Combat Crew Commander, 71st Tactical Missile Squadron, Florennes Air Base, BE
3. July 1988 - April 1990, Ground Launch Cruise Missile Operations Procedure Instructor; Operations Dispersal Instructor, 868th Tactical Missile Training Squadron, Davis-Monthan Air Force Base, Ariz.
4. April 1990 - July 1990, Student, Squadron Officer School, Maxwell Air Force Base, Ala.
5. June 1990 - September 1993, Deputy Crew Commander, Space Surveillance Center; Chief, Space Surveillance Center; Deputy Crew Commander, Space Defense Operations Center, Space Control Operations Division, Operations Directorate, Headquarters, U.S. Space Command, Colorado.
6. September 1993-January 1996, Flight Commander; Chief, Operations Plans Flight; Commander, Operations Support Flight, 4th Space Warning Squadron, Holloman Air Force Base, New Mexico
7. January 1996-June 1996, Student, U.S. Air Force Space Tactics School, Falcon AFB, Colo.
8. June 1996- June 1998 Space Operations Officer, Command and Control Division,

Operations Directorate J-3, Headquarters, U.S. European Command, Stuttgart, Germany
9. June 1998-June 1999 Executive Officer, Operations Directorate J-3, Headquarters, U.S. European Command, Stuttgart, Germany
10. June 1999-June 2000, Student, USAF Air Command and Staff College, Maxwell Air Force Base, Ala.
11. July 2000-October 2000, Student, 392nd Intercontinental Ballistic Missile Training Squadron, Vandenberg Air Force Base, Calif.
12. October 2000-June 2001, Operations Officer, 12th Missile Squadron, Malmstrom Air Force Base, Mont.
13. June 2001-June 2003, Commander, 12th Missile Squadron, Malmstrom Air Force Base, Mont.
14. July 2003-June 2004, Student, National War College, Fort McNair, Washington D.C.
15.June 2004-June 2006, Chief, Space and Missile Defense Operations, Operations Directorate J-3, The Joint Staff, Washington, D.C.
16.June 2006-present, Commander, 460th Operations Group, Buckley Air Force Base, Colorado.

MAJOR AWARDS AND DECORATIONS:
Defense Meritorious Service Medal with Oak Leaf Cluster
Meritorious Service Medal with Oak Leaf Cluster
Joint Service Commendation Medal
Air Force Commendation Medal with Oak Leaf Cluster
Joint Service Achievement Medal with Oak Leaf Cluster
Combat Readiness Medal with Bronze Star
National Defense Service Medal with Bronze Star
Global War on Terrorism Service Medal
1995 USAF Lance P. Sijan Leadership Award

EFFECTIVE DATES OF PROMOTION:
Second Lieutenant August 1985
First Lieutenant August 1987
Captain August 1989
Major August 1997
Lieutenant Colonel May 2001
Colonel March 2006

HENRY F. MAGILL

Colonel Ordnance Corps, US ARMY Retired

Henry F. Magill was graduated from Clemson A & M College, Class 1952 and was awarded his Bachelor's Degree in Textile Engineering. He was very active in the Army ROTC and was an Army ROTC Distinguished Military Graduate recieving his commission as a 2nd Lt USAR. He was also very active in campus activities. Prior to entry in service he was employed with Owens-Corning for a short time. He entered military service on March 7, 1952 and served as an Artillery Mathematics Instructor at the US Military Acadey during 1952 and 1953. During 1953 thru 1954 he served as Commander, Ordance Maintenance Company, 8th Army, Korea. In March 1956 he was appointed Aide de Camp to Major General John B. Medaris, Commander of the Army Ballistic Missile Agency in Huntsvlle, Al. While as Aide de Camp to Medaris he was also an

Adminstrative Aide to Dr Werhner von Braun and gained extensive experience in research, development and testing, and the management of large -scale exploratory and advanced development programs in Ballistic Missile operations and defense. While in service he continued his education and was awarded his Masters Degree in Engineering in 1961 from the Universirty of Alabama in Huntsville, Al. During 1961 thru 1963 he was Project Engineer, Down Range Anti-Missile Measurement Program. During 1963 thru 1966 he was Staff Officer, US Army Research Office in Washington, DC. In 1970 he retired from the Army and later after his retirement he became Chief, Systems Engineering, Advanced Research Projects Agency(ARPA), Saigon, Vietnam during 1970 thru 1971. He became the Director of the Kawajalein Missile Ramge during 1971 thru 1972 and during 1974 thru 1976 he was the Program Manager for the ROLAND All Weather Air Defense Missile System. During 1976 thru 1985 he continued working, supporting Army and Aero-Space industry programs, for Teledyne Brown Engineering, in Huntsville. During 1985 thru 1986 he was the Program Manager, Assault Breaker(Anti-Tank Smart Weapons) for Science and Applications International Corporation(SAIC) in Huntsville, retiring later in 1990 as VP and Operations Manager, Orlando Division of SAIC. After retirement he continued to be associated with the aero-space indusrty in the field of systems engineering and program managment with Coleman Research in Orlando as VP retiring from them in 1993.

Magill was interested early in aviation and after he soloed his first aircraft in 1945 he later recieved his Private Pilots license. When stationed in Huntsville in the 50's he flew a Taylorcraft aircraft and was part owner. He later stopped flying in the 50's after his partner in the Army was transfered to White Sands and took the Taylorcraft with him.

Col Magill during his military service attended and graduated from the following:

US Defense Managment College, Ft Belvour, VA.
US War College , Carlisle Barracks, PA.
US Command and General Staff College, Ft. Leavenworth, KS
US Army Ordance School, Aberdeen Proving Frounds, MD
US Army Artillery School, Ft. Sill, OK.

Some of Henry's military decorations include the following:

Bronze Star
Army Commemdation Medal (2 Oak Leaf Clusters)
Legion of Merit
Korean Service Medal
National Defense Service Medal (1 Oak Leaf Cluster)
Army Occupation Medal
Meritorious Service Medal (1Oak Leaf Cluster)
Joint Service Commendation Medal
Vietnam Campaign Medal
Vietnam Armed Force Honor Medal(First Class)
Vietnam Cross of Gallantry w/Palm(Unit)
OSD Identification Badge
Meritorious Unit Citation (1 Oak Leaf Cluster)

Henry has been an Adjunct Instructor, Program Managment, Southeastern Institute of Techonology in Huntsville, Al during 198 through 1986 and Guest Lecturer in South Africa in the 80's and in Egypt during 1984.

He is now retired but continues to be a Consultant -Program Management and Proposal Development

JACK C. SPILLERS
Col USAFR-Ret

Jack C. Spillers was born October 4, 1931 in Greenville County as the oldest son of Reverend and Mrs. J.W. Spillers. He soloed a Piper J-3 Cub on his 16th birthday after four and a half hours of dual flight instruction and was graduated from Clinton High School in 1949. He then enrolled in Clemson A & M College (now Clemson University) and graduated in 1964 with a Bachelor of Science in Textile Engineering and received his commission as a 2nd Lt in the United States Air Force thru the AFROTC program. While at Clemson he was a member of the Senior Platoon. He entered active duty with the USAF in March 1954 and after completing his pilot training at Webb AFB, Texas, he received his USAF Pilot Wings with the Class 55L. He then went to Craig AFB, Al for basic flight instructor school, remaining there for 3 years instructing in North American T-28's and Lockheed T-33. In September 1958 he was assigned to Boeing B-47 Flight training at McConnell AFB, Kansas and subsequently, went to Mountain Home AFB, ID as an Aircraft Commander and Instructor pilot. He attended Squadron Officers School in 1960 and the Air Command and Staff College at Maxwell AFB during 1965-1966. He then went to Seymour Johnson AFB, NC where he completed the F-105 RTU in 1966.

He was assigned to the 388th Tactical Fighter Wing, 469th Tactical Fighter Squadron at Korat RTAFB in 1966 and flew 114 combat missions over North Vietnam. He was shot down by ground fire in North Vietnam on his 46th combat mission which occurred on Easter Sunday, 1967, and he was later rescued by the Jolly Green Giants from the sea after he had ejected from his damaged F-105. He completed his combat tour in August 1967.

He was then assigned to HQ. USAF in the Pentagon in Washington, DC as an Action Officer in Strategic Future Force Development. In November 1969, he voluntarily returned to F-105's at Takhli RTAFB, Thailand with the 355th Tactical Fighter Wing where he became the Operations Officer and later as a Lt Col he became the Commander of the 357th Tactical Fighter Squadron. He flew, during his second tour , an additional 106 combat missions of which several were as Mission Commander of strikes into North Vietnam.

In November 1970 he was assigned to Hq 9th Air Force at Shaw AFB, SC as Chief of Flight Safety and promoted to Col in March 1973. He then attended the Air War College during 1973-1974 and simultaneously obtained an MS degree in Counseling and Guidance from Troy State University, Troy, Al. Col Spillers was then assigned to the 23rd Fighter Wing at England AFB, La. as Director of Operations, flying A-7D's and one year later he became Wing Vice Commander.

In 1976 he was assigned to Hq. USAFE at Ramstein, Germany as Director of Plans. In 1978 he became the Senior Liaison Officer to Hq USAREUR in Heidleberg, Germany. In 1980 he was returned to the USA to Shaw AFB, SC as the 9th AF Director of Tactical Operations. He retired from the USAF on April1, 1984.

During his 30 year military career he flew a number of aircraft, these being the PA-18, AT-6, T-28, T-33, B-47, T-39, F-105, and the A-7D.

Military Decorations:

Silver Star
Two Legions of Merit
Four Distinguished Flying Crosses
Bronze Star
30 Air Medals
Purple Heart
Two Meritorius Service Medals

AF Commendation Medal
RVN Gallantry Cross w/device
Vietnam Campaign Medal
Vietnam Service Medal
Combat Readiness Medal
National Defence Medal

Col Spillers married the former Jean Wilison of Laurens, SC on April 2, 1953 and they were the parents of three children, Carol, Janice, and James. His first wife died on August 12, 2003 and he later in 2005 married Patrica Ann Gilespie and became the father to her four children. They reside in Sumter, SC and she is a 6th grade teacher at the HillCrest Middle School. They have a total of 13 grandchildren. He is an active member of the First Baptist Church here in Sumter.

After his Air Force retirement he earned an Associate Degree in Environmental Control Technology and founded his company Jacks Heat Pump Service Inc there in Sumter and continued in business there till he entered full retirement in 1994.

Professional:

Life Member and Past Captain of the Order of Daedallians, Chaplain of the Military Officers Association of America, Shaw Chapter at Shaw AFB, Life Member of Air Force Association, American Legion, Life Member of the Red River Valley Fighter Association, and a Member of the Sumter Radio Control Model Airplane Club.

COLONEL JOHN M. McCAIN

Col John M. McCain is the Chief of the Gulf Cooperation Council Division, Regional Affairs Directorate, Deputy Under Secretary of the Air Force for International Affairs, Headquarters U.S. Air Force, Washington, D.C. The mission of SAF/IARG is to provide Air Force policy oversight and guidance to international programs supporting national security objectives through politico-military affairs and security assistance programs.

Col McCain graduated in 1983 from Clemson University and was commissioned a second lieutenant through the AFROTC where he earned Distinguished Graduate honors. After completing Undergraduate Navigator Training at Mather AFB, CA, he was assigned to McConnell AFB, KS, in 1985, as a KC-135A and KC-135R navigator. During his tour at McConnell, Col McCain served as an instructor navigator and standardization evaluation navigator specialty chief. In 1990, Col McCain was assigned to the 93d Air Refueling Squadron, Castle AFB, CA, as an instructor in Strategic Air Command's selectively manned Combat Crew Training School.

In 1994, Col McCain was assigned to Headquarters Air Mobility Command, Scott AFB, IL as an Operational Programming Officer in the Plans and Programs Directorate. In 1997, Col McCain was assigned to the 344th Air Refueling Squadron, McConnell AFB, KS as the Assistant Operations Officer in one of Air Mobility Command's most heavily tasked KC-135 units. During this tour, Col McCain deployed to the Kingdom of Saudi Arabia supporting Operation SOUTHERN WATCH and was the deployed commander of the 22d Air Refueling Wing's Red Flag 98-3 detachment at Nellis AFB, NV. In 1998, Col McCain was selected as Chief, 22nd Operations Group Training where he ensured the mission ready status of over 500 crew members in Air Mobility Command's most diverse KC-135 wing. Later that year, Col McCain was selected as the Senior Executive Officer to the 22d Air Refueling Wing Commander and in early 1999 Col McCain was assigned as Director of Operations for the 22d Operations Support Squadron.

Col McCain assumed command of the 36th Student Squadron, Maxwell AFB, AL

Chapter I
1927-1938

Preston Mood, Nat Green,
Norwood Mood, Professor John Marshall,
Dallas Sherman

Wing and Frame of Little 372
before being covered

George M Kieghtly
Greenville Historical Society

Kieghtly's Ford Tri-Motor NC7865
Greenville Historical Society

Lawrancc Engine C-3

A Typical Heath Aircraft

110 HP Rhone Engine

Dallas Sherman and Little 372

Dallas Sherman in Little 372

Theodore Boscelli WWII Navigator for Pathfinder Flight for President Roosevelt to Yalta TAPS

Chapter 2

1929-1934

A Typical 1929 Primary Glider

Aero Club Aircrafts- Little 372, Glider G1, and Eagle Rock Aircraft

Jim Nesbit and his
Bonanza Aircraft

Nesbit's Airport USGS Photo

Aero Plantation

Jim Propping Aeronca

Tom H. Davis's Taylorcraft E2 and
Piedmont 737

Liberty Engine
US Air Force Museum

Chapter 3
1933-1944

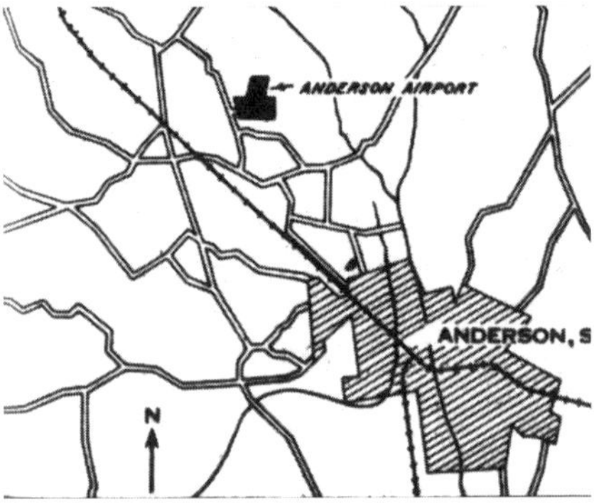

1935 Anderson SC Airport - Air Register

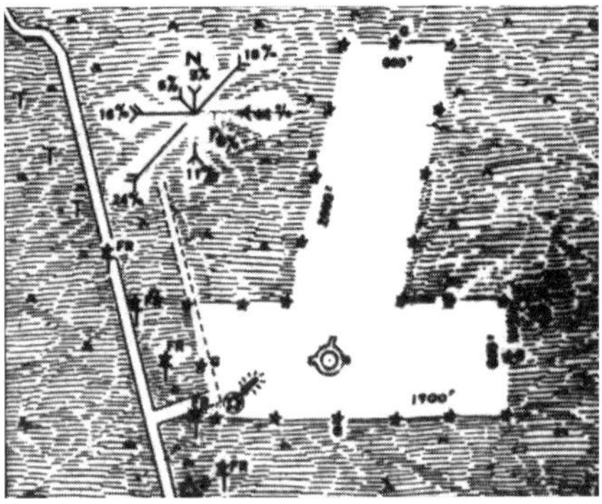

Anderson, SC Airport - Air Register

Flying Cadets 1941 TAPS

Flying Cadets 1941 TAPS

Flying Cadets 1942 TAPS

Flying Cadets 1942 TAPS

Flying Cadets 1943 TAPS

Flying Cadet Watkins and Piper Cub
Watkins Photo

1941 Clemson Flying Cadets
Watkins Photo

Army Specialized Training Patch

1944 ASTP Clemson's Group 1
Clemson-ASPT "Saturated" Yearbook

1944 ASTP Clemson's Group 2
Clemson-ASPT "Saturated" Yearbook

CPTP Stinson Watkins Photo

CPTP Stinson Watkins Photo

Clemson Flying Cadet Wings Watkins Photo

CPTP Aircraft Watkins Photo

Department of Commerce

This Is To Certify: That Miles Hunter Watkins

has satisfactorily completed the following controlled courses of instruction for

Private Pilot

GROUND COURSE

History	Parachutes	Civil air regulations	Theory of flight	Instruments
Meteorology	Aircraft	Practical air navigation	Engines	Radio

FLIGHT COURSE

Taxiing	Take-offs	Forced landings	Steep power turns	Power landings
Air work	Landings	Precision landings	Slips—forward, side	Dragging areas
Stalls	Spins	Figure eights	Power approaches	Cross country

and is accorded all the rights and privileges of that grade as evidenced by Certificate of Competency No. 93416-41

Issued this Fifteenth day of September 1940

Approved:

The Clemson Agricultural College

Assistant Secretary of Commerce

Anderson Flying Service

Watkins CPTP Private Pilot Course

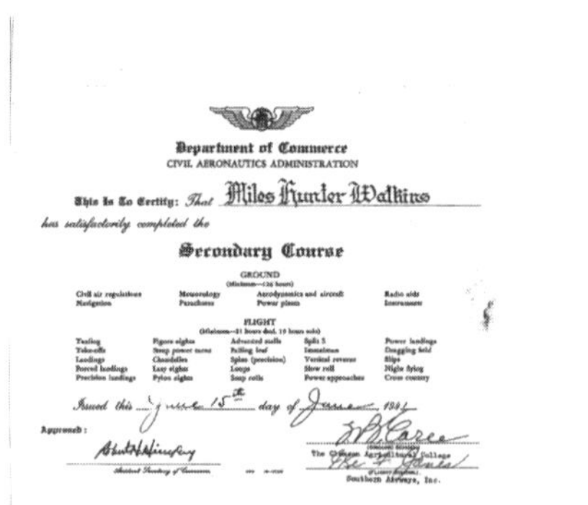

Department of Commerce
CIVIL AERONAUTICS ADMINISTRATION

This Is To Certify: That Miles Hunter Watkins

has satisfactorily completed the

Secondary Course

GROUND

Civil air regulations	Meteorology	Aerodynamics and aircraft	Radio aids
Navigation	Parachutes	Power plants	Instruments

FLIGHT

Taxiing	Figure eights	Advanced stalls	Split S	Power landings
Take-offs	Steep power turns	Falling leaf	Immelman	Dragging field
Landings	Chandelles	Spins (precision)	Vertical reverse	Slips
Forced landings	Lazy eights	Loops	Slow roll	Night flying
Precision landings	Pylon eights	Snap rolls	Power approaches	Cross country

Issued this June 15th day of June 1941

Approved:

Assistant Secretary of Commerce

The Clemson Agricultural College

Southern Airways, Inc.

Watkins CPTP Secondary Course

Clemson Flying Cadets
Hufford Collection

Clemson Flying Cadets
Hufford Collection

WWII Army Air Corps Cadets Assemble
on the Quadrangle
TAPS

Clemson Army Air Corps Cadets
Marching to Classes
TAPS

Chapter 4
1945-1960

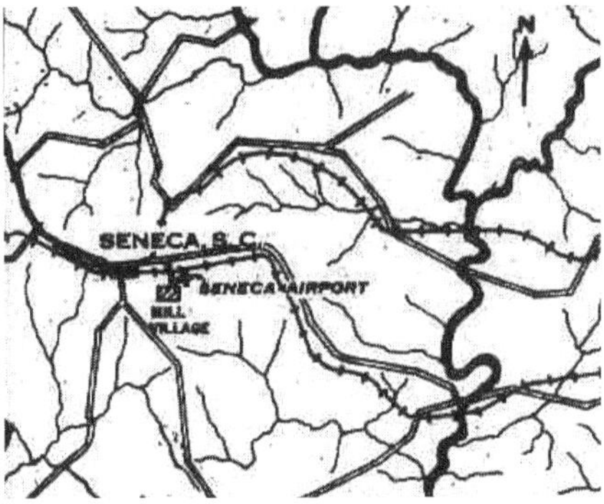

1935 Seneca Airport
Air Pilots Registery

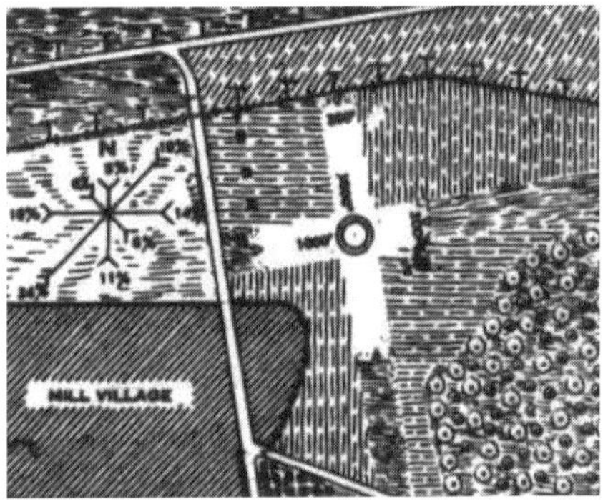

1935 Seneca Airport Closeup
Air Pilots Registery

Jenny Aircraft at the
Seneca Tribble Field Airport

Clemson College 1947 and Veterans
Housing Herb Morgan Photo

Begining of Site Preperation for
Clemson-Central Airport

1947 Clemson-Central Airport
Aerial View

Joe Salvadore (Co Owner of Airport) and his daughter Dottie in front of Hanger

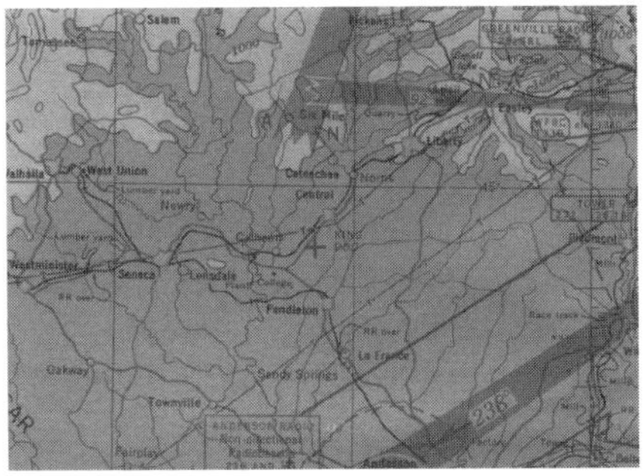

1948 Charlotte Sectional Showing Clemson-Central Airport

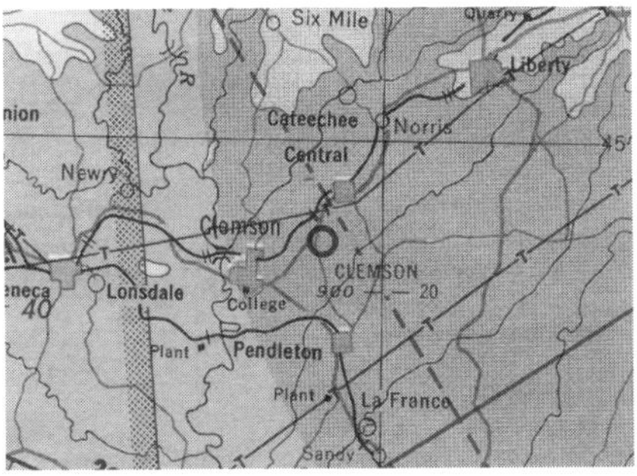

1948 Charlotte Sectional Showing Clemson-Central Airport

Leonard Meldeau first Aircraft

Leonard, Me, and L-3 Aeronca

Leonard, Me, Mother, Dad
My 1946 Solo Picture

Lt Meldeau and his Spitfire in N Africa

Meldeau Flying Service Poster

Leonard and Margaret Meldeau

1948 Clemson Aero Club

1949 Clemson Aero Club

1949 Aero Club TAPS

1950 Aero Club members

1950 Missouri Clemson Aero Club at
Columbia, Mo.
J. Rostron

Aero Club 1951
TAPS

Ivan's Anderson's Clemson
Aero Club Piper J-3 Aircraft N41026
Refurbished and Flying

Johnny J. Weeks

Knight Twister 711V KT

Johnny J. Weeks Super Cub

Johnny J. Weeks Seaplane

Johnny J. Weeks Cessna 210

Johnny J. Weeks Culver Cadet

Johnny Week's Knight Twister Refurbished 2003 By Fred J. Ring and Tom Rowden of Kendrick Idaho

Johnny Week's Knight Twister Refurbished 2003 By Fred J. Ring and Tom Rowden of Kendrick Idaho

My Mom, her dog, and Ercoupe

My Ryan Solo Photo

Ryan and Stearman at the Clemson-Central Airport

William F. Wilson Flight Instructor for Meldeau Flying Service TAPS

1951 Military Staff (Army and AFROTC) TAPS

1953 AFROTC Staff TAPS

1946 Air ROTC Shoulder Patch

Today's Clemson's AFROTC
Detachment 770 Patch

Chapter 5
1960-1963

1957 Cherry Bottom Area before
Airfield Construction

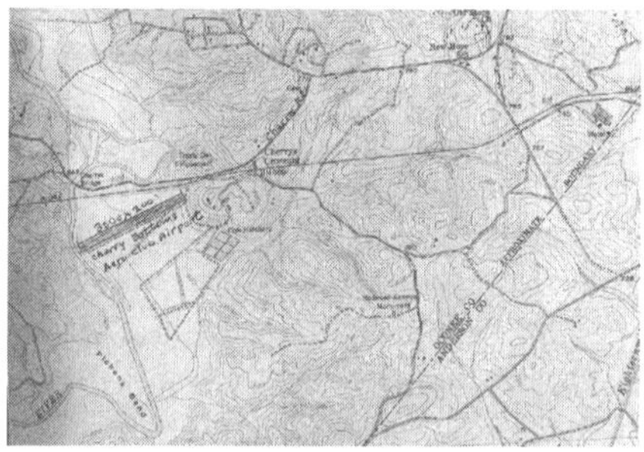

Proposed Cherry Bottom Airport

Danielio Lazzilo in Bonanza

Danielio's Bonanza

Danielio's Pacer

Danielio's Cessna

Danielio's Cessna 337

Clemson Aero Club Offers Flying Lessons to Students

Membership in the Clemson Aero Club, Inc., is open to all students, to the members of the faculty, and to college personnel. It is the purpose of the club to promote interest in flying among the people of Clemson College.

The Aero Club offers a ground school to all members and guarantees each member that he will pass the Federal Aviation Written Exam. Members can earn their Student's license, Private Pilot's license, Commercial license, and Instrument Pilot's license.

The club has an F.A.A. rated flight instructor who instructs the members and conducts the ground school. Each member is covered by a one million dollar insurance policy which the club provides for them while they are flying in the club planes.

1963 Clemson Aero Club TAPS

Local Aero Club Promotes Interest in Aviation

1964 Clemson Aero Club TAPS

1964 Clemson Aero Club Taylorcraft TAPS

1964-1965 CAC Patch

Aeronca L-3 Destroyed in Crash at
Lake Hartwell Near Clemson

1961 Angel Flight Ladies TAPS

Clemson Aero Club's Original N95809
Now Owned by Ron Portzer of Jasper, AL

Danielio's Award Photo

Danielio's Cruz de la FAV

Ed Byar's Swift N3727K

Ed Byars and his Fairchild N48320

Funk B-1

Mike Smith, his son, and Mike's Jet

Chapter 6

1945-1960

Disassembled Little 372 On Trailer For Transfer to SC Museum

Disassembled Little 372 On Trailer For Transfer to SC Museum

372 in SC Museum

1979 Clemson Aero Club

Jerry Wigington Buckner Aircraft

Jerry Wigington Stearman PT-17 Aircraft

Clemson Oconee Airport 1979

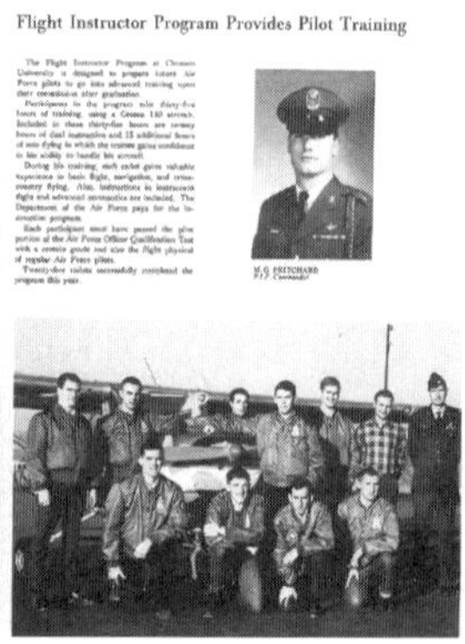

Flight Instructor Program Provides Pilot Training

1968 AFROTC Flight Instructor Program
at Clemson Oconee Airport
TAPS

1968 AFROTC Flight Instructor Program
at Clemson Oconee Airport
TAPS

1969 AFROTC Flight Instructor Program
at Clemson Oconee Airport
TAPS

CHAPTER 7

Caroline Burley Miller

Caroline Burley Miller in Early Clemson Aircraft

Clemson Athletics Aircraft
Operations Building

Clemson Athletics Aircraft 1

Clemson Athletics Aircraft 2

1972 Flying Club's Cessna 150 TAPS

1983 Clemson Aeronautical Association TAPS

Aeronautical Association 1984 TAPS

Flying Club Aircraft Plane 1980

Flying Club's Aircraft on Display on Campus 1986 TAPS

1980 Aeronca NC4455E at
Pickens County Airport

Don Fraiser Refurbished Aeronca
NC4455E

A Typical Aircraft Flown By Eric Prufer

Chapter 8

Andy Applegate Airline First Officer

Chip Houck and Karen Schwartz in Clemson Aeronautical Association Cessna 150 1981

Aero Club Members Washing Aircraft 1989

Dallas Sherman's Tiger Aircraft Gift

Clemson Aero Club's Tiger Refurbished N9526L and Still Flying

A Typical AirTran Boeing 717 Flown By Jeff Treece

Scott Bowen Thunderbird Slot Pilot and Aircraft

1979 Clemson Aero Club Logo

1991 Flying Club Members TAPS

1994 Flying Club Members TAPS

1996 Flying Club Members TAPS

Cessna 150 N65407 Tim Miller

Clemson Airport 1989 Tim Miller

Death Valley Before the game

Clemson's Death Valley and Football
Tim Miller

Flying Club Members 1989-1990

George Todd's Aproach to Rutherford County Airport, NC

Clemson Aero Club Cessna 150 at Rutherford County Airport, NC

737RP at Rutherford County Airport, NC

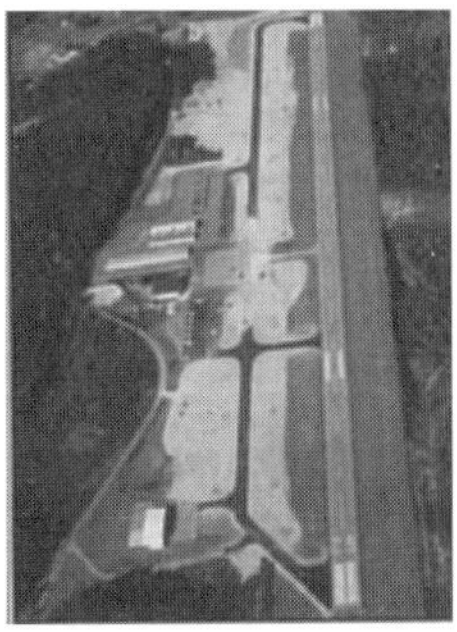

KCEU Aerial RW 270

Some Aircraft and a Glider on the CEU ramp

Tim Miller and N737RP

Eagle Ridge Airport, SC USGS photo

Flying Club 172
Now Owned By Cookie

N65407 in Front of Campus Building

Organization Day 2004 Aircraft
Taxiing Down Highway 93

N65407 Taxiing Down the Highway

2000 Flying Club TAPS

Club Aircraft Parked In Front of the
Clemson Library

Cookie Fitch's Cessna N2650L

Doyle Graham's Solo Day

Chapter 12

Clemson Oconee Airport Commissioners
2006

Clemson Parachute Club Dixie Devils
Jumper Away

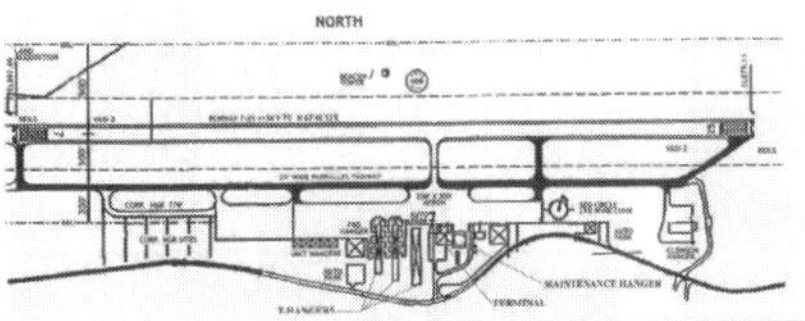

KCEU Airport Diagram

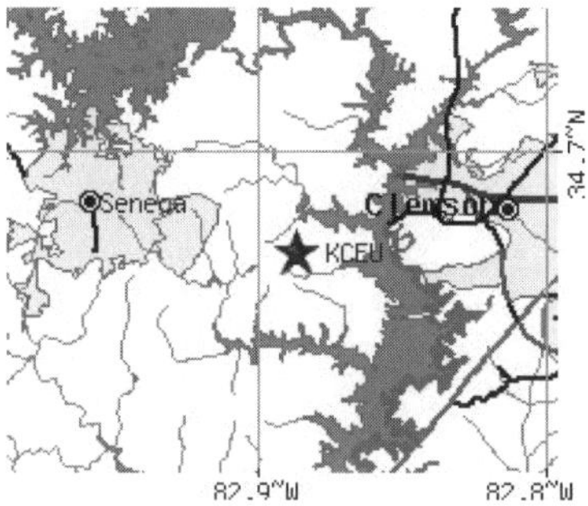

KCEU Map

Chapter 13

Bob Russell and his Legend Aircraft
Scott Krause Photo

Bob Russell's Chipmunk
Joseph K. Newsom Sr. Photo

Bob Russell and his Legend Aircraft
Scott Krause Photo

Lt Edward A. Smith and His F-86A Sabre
in Korea on Alert Pad
John Starr Photo

Jim Woodfin F-86D Sabre Pilot -
all weather Interceptor

Cliff McClure in his Manhigh 3
Stratospheric Balloon Flight Gondolla

Rudolph Anderson U-2 Pilot Shot Down
During Cuban Missile Crisis

U-2A Aircraft

1967 Col Jack C. Spillers

1970 Col Jack C. Spillers

Otha H Vaughan Jr in KC 135
Zero-G-Flight

Joe Rostron and his 1946 Cessna 140

Stealth Fighter Nighthawk F-117

Thunderbolt A-10

Walter Smithwick's F-101 Voodoo Photo
Recon Aircraft at Udon

Captain E. Shelly in
Civil Air Patrol Uniform

THE CLEMSON CORPS

Clemson Corps Logo

in June 2000 and in Aug 2001 was selected Vice Commandant of the Air and Space Basic Course, the USAF's largest resident officer PME course. In 2003 Col McCain was assigned to the Deputy Under Secretary of the Air Force International Affairs, Mideast/ Africa Division as the Country Director for Jordan and Iraq. Col McCain assumed his current position in July 2005.

EDUCATION:

2000 Air War College (seminar)
1996 Air Command and Staff College (seminar)
1993 MBA, Embry-Riddle Aeronautical University, FL
1989 Squadron Officer School, Maxwell AFB, AL
1983 BS, Administrative Management, Clemson University, SC

ASSIGNMENTS:

1. 1984, Student Undergraduate Navigator Training, Mather AFB, CA
2. 1984-1985, Student KC-135 Combat Crew Training School, Castle AFB, CA
3. 1985-1987, Navigator KC-135 91st Air Refueling Squadron, McConnell AFB, KS
4. 1987-1989, Instructor Navigator KC-135 384th Air Refueling Squadron, McConnel AFB, KS
5. 1989-1990, Stan/Eval Navigator 384th Bomb Wing, McConnell AFB, KS
6. 1990, Stan/Eval Navigator Speciality Chief 384th Bomb Wing, McConnell AFB, KS
7. 1990-1991, Instructor Navigator KC-135, Combat Crew Training School 93rd ARS, Castle AFB, CA
8. 1991-1992, Assistant Flight Commander Combat Crew Training School, Castle AFB, CA
9. 1992, Academic Instructor Combat Crew Training School, 330 Flying Training Sq, Castle AFB, CA
10. 1992-1993, Instructor Nav Combat Flight Instructor Course, 330 Flying Training Sq, Castle AFB, CA
11. 1993-1994, Chief of Quality Improvement, 398 Operations Group, Castle AFB, CA
12. 1994-1997, Operational Programming Officer, HQ Air Mobility Command, Scott AFB, IL
13. 1997-1998, Assistant Operations Officer, 344th Air Refueling Squadron, McConnell AFB, KS
14. 1998-1999, Chief of Training, 22nd Operations Group, McConnell AFB, KS
15. 1999, Wing Executive Officer, 22nd Air Refueling Wing, McConnell AFB, KS
16. 1999-2000, Director of Operations, 22nd Operations Support Squadron, McConnell AFB, KS
17. 2000-2001, Commander 36th Student Squadron, Air and Space Basic Course, Maxwell AFB, AL
18. 2001-2003, Vice Commandant, Air and Space Basic Course, Maxwell AFB, AL
19. 2003-2005, Country Director, Deputy Under Secretary of the Air Force International Affairs
20. 2005-2006, Chief, Gulf Cooperation Council Division, Deputy Under Secretary of the Air Force International Affairs
21. 2006-Present, Student, Air War College, Maxwell AFB, AL

FLIGHT INFORMATION:

Rating: Master Navigator

Flight hours: 2,700
Aircraft flown: KC-135A, R

MAJOR AWARDS AND DECORATIONS:
Meritorious Service Medal with four oak leaf clusters
Air Force Commendation Medal with one oak leaf cluster
Air Force Achievement Medal with three oak leaf clusters
Combat Readiness Medal with three oak leaf clusters
Distinguished Graduate - Air Force ROTC, Clemson University
2003 Lance P. Sijan Leadership Award Sr. Officer Category - Air University, Maxwell AFB AL
2004 Action Officer of the Year-Deputy Under Secretary of the Air Force International Affairs

EFFECTIVE DATES OF PROMOTION:
Colonel Sep 1, 2005
Lieutenant Colonel Nov 1, 1999
Major Jun 1, 1995
Captain Jul 28, 1987
First Lieutenant Jul 28, 1985
Second Lieutenant May 13, 1983

MAJOR RUDOLPH ANDERSON

U-2 Reconnaissance Pilot-Cuban Missile Crisis

Rudolph Anderson was graduated from Greenville High School in 1944 and in September he was accepted to enter Clemson A & M College where he majored in Textile Engineering. While at Clemson he was a member of the Army ROTC program and he became a member of the Senior Platoon Drill Team. He was active in inter-mural sports as well as community leagues. After graduation he worked for a while but decided to enter the USAF. After further training he was assigned to the 15th Reconnaissance Squadron in Japan flying the RF-86F aircraft. While in Japan he participated in a number of reconnaissance flights that were carried out over China and Russia. For these flights during the early period of the Cold War he was awarded two Distinguished Flying Crosses. After his service in Japan he was returned to the United States and was later assigned to the 4080th Strategic Reconnaissance Wing at Laughlin AFB, Texas. While in this Wing he was selected to begin training to learn to fly the new super secret reconnaissance aircraft called the U-2 aircraft that was being developed to conduct strategic reconnaissance over enemy military installations. Only the best pilots were selected in the early phases of its operation and those pilots that were military trained pilots and reserve officers were recruited by the Central Intelligence Agency (CIA) to fly this aircraft. After the initial production aircraft were delivered to the CIA for their use then additional aircraft were assigned to the USAF for their use. Rudy was assigned in this new class of Strategic Air Command pilots in 1957 that were selected to be trained to fly these special aircraft. The U-2 pilots were trained at the Groom Lake, Nevada auxillary field called the ranch where the U-2 flight development program was being conducted by Lockheed pilots and the CIA. The early aircraft missions were the High Altitude Sampling Missions (HASP). These missions were designed and the aircraft were modified with sampling devices and were flown to collect samples of the radioactive debris that occurs in the upper atmosphere after a nuclear device is detonated. These flight were used to check on the status of nuclear developments of Russia, Chinese, and the French Nuclear Weapons programs. Rudy flew some HASP missions during the end of March 1957 from Eielson AFB, Alaska.

During early 1962 CIA flights were conducted over the Island of Cuba and obtained photo evidence of a buildup of Intermediate Range Ballistic Missile (IRBM) launch facilities. This was a system that once in place could strike centers in to the USA so an intensive surveillance program was needed to be initiated. Richard Heyser and Rudy were sent to the Groom Lake airfield and were trained to fly the CIA aircraft version, a different version that they had been previously trained to fly. These aircraft U-2C's were equipped with better cameras and other instrumentation than were available in the USAF U-2A aircraft. After their completion of their training, Heyser flew the first USAF surveillance over-flight mission from Edwards AFB to Cuba and after his mission was accomplished he landed at Mc Coy AFB where the 4080th SRW was now deployed to McCoy AFB, Fl to begain their surveillance of Cuba. A number of these type mission were flown over the next few days and the Cuban Missile Crisis began to develop. It was on one of these surveillance missions on October 26, 1962 that Rudy's pressure suit, that protects the pilot from the environment above 60,000 feet and allows hin to operate the aircraft in this environment, was damaged by a metal shard from an exploding SA-2 missile below his U-2. By losing the protection from the environment he perished and the aircraft crashed. He was first awarded posthumously the Distinguish Service Medal for his heroic action. Since we were in the Cold War period at this time it was the highest award that was authorized. Later he was awarded the First Air Force Cross, the Purple Heart, and the Cheney Humanitarian Award.

On October 28, 1962 Nikta Kruschev and John F. Kennedy reached an agreement whereby that Russia would remove all offensive missiles and other munitions from Cuba, the Cuban Missile Crisis was over and the world had been saved from a nuclear exchange between the super powers at that time.

LT COLONEL JAMES BLACK

U-2 Cold War Pilot

James Black entered Clemson first in the spring of 1944 to enter summer school to attend two sessions prior to the September fall semester so that he could get three semesters before February, 1945. By April he would become 18 and usually three weeks before your 18th birthday in York county you would be on a bus reporting to Ft. Jackson for a physical. He didn't want to be in the Army (Infantry) so by mid April he decided to join the Navy. He had completed three years of his education and with a fourth semester he then would be eligible for Naval Flight school. After he signed up he was assigned to attend small Emenry and Henry College in Bristol, Va the Navy somehow made a mistake in that 40 people were assigned but the College could take only 20 people so he and the 19 were sent to a small Catholic College in Winona, Minnesota to continue their preflight education.

While at Winona he decided he would learn to fly by paying for his lessons and soloed a J-3 Piper Cub. V-J Day came in the middle of August so the Navy let them finish their education and shipped Jim off to the Great Lakes Training Center for reassignment. He was then assigned to a sea going Tug ATR-82 that was usually based at Yerba Buena Island, between San Francisco and Oakland, CA. He remained there till he was discharged in August 1945. After his release from service he re-entered Clemson. While at Clemson he majored in Textile Manufacturing and was awarded his degree, Class of '49. He then went to work for his father in the family business, a Textile-Mill Hardware Company in Clinton SC where he worked till late 1950. He initiated paperwork to enlist in the USAF as an Aviation Cadet. However after waiting for a while he finally received his orders assigning him to Class-53B, reporting for duty on in February 1952 at Bainbridge, GA. There he spent a month in preflight and six months of primary pilot training learning to fly the

North American AT-6G aircraft. He completed his course work and flying and was then assigned to Larado AFB, Texas where he continued his training and flew the T-28 for the next three months and then three more months of training and flying the T33.

His flight instructor asked him what his preference for further training might be. Jim was not really sure so the instructor suggested that he might want to be a Reconnaissance Pilot. After Jim received his commission he was assigned to Moody AFB, for the next six weeks of Instrument Training and then was assigned to 18th Tactical Recon Squadron, Shaw AFB, Sumter, SC for three and a half months of reconnaissance training. Shortly after getting to Shaw the Korean War ended and he was ordered to report to K-14, (Kimpo AB, Korea) assigned to the 15th Tactical Reconn Squadron where he remained for some 20 months (Oct 53 thru Sept 55.) He flew a number of special reconnaissance missions flying the RF-86Faircraft throughout the Far East.

During his military flying career Jim flew the AT-6G, T-28, T-33, RF-80, FR-86, RF-86F, U-3, U-2, and the T-39, amassing over 4471 hours. He was awarded the Distinguish Flying Cross, the Air Medal with one Oak Leaf Cluster, and the Bronze Star during his USAF career.

After tours of duty in California, Hawaii, and Viet Nam he was returned to Shaw AFB, Sumter, SC where he retired from military service. He was employed as a commercial pilot for Beaukinit Corporation, a Textile Company in Raleigh, NC. He later flew again as a corporate pilot for a Highway Road Project Construction Company till he retired again and then moved to Cary, NC.

LT CLIFTON M. "DEMI" MCCLURE

High Altitude Balloon ResearchPilot-ManHigh III

A native of Anderson, SC he was the son of the late Clifton M. McClure. and the late Frances Allen McClure. McClure was known to his friends as "Demi". He got that name by being born on election day, November 8, 1932, the date that Franklin Roosevelt, a Democrat, was elected to become the President of the United States. The doctor who delivered him, held him up, spanked him, and said "the first little Democrat," gave him that name. Cliff was taught by his father while in high school to fly a Piper Cub and together they rebuilt a Piper Criuser. Cliff loved flying both aircraft and sailplanes flying.

While at Clemson he was very active in the AFROTC program, a member of the Pershing Rifles and the Senior Platoon, both precision drill teams. He attained the rank of Cadet Colonel, the Commander of the Cadet Corps. Upon graduation in 1954 he was commissioned a 2nd Lt USAFR and received a degree in Ceramic Engineering with a minor in Electrical Engineering and later in 1955 his Master's degree in Materials Engineering. Later in 1955 he entered the Air Force to attend USAF Pilot Training. After completing USAF pilot training he served as a Jet instructor pilot training young pilots in the T-33 at Bryan AFB. While at Bryan he attended Texas A & M and studied Astronomy and Astrophysics.

His professor at A & M, being aware of the Col John Stapp's High Altitude Manned Space Cabin Research Program being conducted at Holloman AFB, contacted Col Stapp and recommended that he take a look at the young 2nd Lt's qualifications as a Research Pilot.

Col Stapp talked with McClure and shortly thereafter he had him transfered to the USAF Aeromedical Laboratory at Holloman ATB to be a member of his research staff investigating the effects the Space Environment related to possible future manned space flights.

While at Holloman he participated in the early USAF Zero Gravity Research program, flying as a pilot and test subject in the F-94C, and later he was selected to be a member of the High Altitude research group to learn more about the effects of space travel and its stress on human beings using subjects flying in high altitude balloon flights. He was selected as the pilot for the Manhigh III balloon flight and was the third pilot to fly a helium balloon to an altitude of over 99,700 feet on October 8, 1958 and the third person to see the curvature of the earth from space. In September 1960 he was awarded the Distinguished Flying Cross by the USAF for his Record Breaking high altitude balloon flight and his contributions to the understanding of the effects of space environment on humans.

Col. David Simons in his book "ManHigh" talked about the significance of McClure's flight as follows: "The lesson of this unexpected but priceless experiment was abundantly clear to all of us. An essential quality of necessary to an astronaut would be stamina, not in a purely physical sense but in a psycho-physical sense: a combination of deep physical reserves plus the all-important emotional determination to use those reserves..... Its profound importance was inescapable when we look at the record of McClure's performance.... without communication he could get no help or comfort from the ground. He was completely on his own and he knew it. But instead of flipping into panic, his determination stiffened...By some factor of fate McClure had been dropped into the right place at the right time to be selected for Man-High..." Upon leaving the Air Force, he joined the South Carolina Air National Guard where he flew the F-104 fighter aircraft with the 157th Fighter Interceptor Squadron at McEntire Air Force Base near Columbia, SC. His unit was activated in 1962 and were sent to Spain as a result of the Berlin Crisis. He continued to remain in the Guard for a number of years. After his release from the guard he applied for and received a Fellowship at Mellon Research, using his Materials Engineering degree researching corrosion of metals. In 1967 he moved to Huntsville, Al. where he worked as a contractor to NASA Marshall Space Flight Center (MSFC). He was assigned to the group that designed the first large Neutral Buoyancy Facility in which astronauts could be trained to work in the simulated weightless environment of space. He determined the type of underwater training that should be used by the Neutral Buoyancy program by becoming himself certified under several of the SCUBA training programs (NAUI, YMCA, and NASDS.)

In addition to his consulting work with NASA MSFC and other space related industries he opened a dive shop and built his own diving facility to train his students using the self contained diving gear. In 1973 he and his wife Bobbie formed their own company Consumer Fuels (Incorporated in 1979) initially for the purpose of providing natural gas for use in motor vehicles. The company evolved developing breathing air systems for hazardous environments and for a number of years supplied them to major companies and government facilities that were involved in asbestos removal, nuclear waste control, and hazardous waste disposal.

McClure was inducted into the International Space Hall of Fame at Alamogordo, NM in 2001.

He continued to operate his business with his wife Bobby until his untimely death on January 14, 2000.

LT COLONEL SCOTT C. BOWEN
Thunderbird Pilot

Lieutenant Colonel Scott C. Bowen is currently the Operations Officer for the 52nd Operations Support Squadron, Spangdahlem Air Base, Germany. The 52d Operations Support Squadron supports the combat readiness of three fighter squadrons and one air control squadron; and is committed to NATO and US national taskings by providing scheduling, pilot training, plans, weapons and tactics, intelligence, weather, airfield management, and air traffic control. The OSS directs a $55 million annual flying hour program for 60 aircraft that the base possesses. It also leads life support and resource management operations for over 110 fighter pilots.

Lieutenant Colonel Bowen was born in Munich, Germany. In 1988 he graduated from Clemson University, receiving a Bachelor of Science degree in Computer Science and receiving his commission through the AF Reserve Officer Training Corps as a distinguished graduate. He completed Undergraduate Pilot Training at Columbus AFB, Mississippi in 1989 as a distinguished graduate, and F-16 training at MacDill AFB, Florida in 1990 as a distinguished graduate. He has served at various assignments including Holloman AFB, New Mexico; Luke AFB, Arizona; Nellis AFB, Nevada; Maxwell AFB, Alabama; Larissa, Greece and Spangdahlem AB, Germany. He is a command pilot with more than 2,800 hours in the F-16 aircraft.

Lieutenant Colonel Bowen is married to the former Sonja Lydia Hienz of Dudeldorf, Germany. They have a 5-year old daughter named Hannah Marie.

EDUCATION:

1988 Bachelor of Science Degree, Computer Science, Clemson University, SC
1995 Squadron Officer School (residence), Distinguished Graduate, Maxwell AFB, AL
1998 Master of Aeronautical Science, Embry Riddle Aeronautical University, Luke AFB Campus
2002 Master of Operational Art, Air Command and Staff College (residence), Maxwell AFB, AL
2004 Joint and Combined Warfighting School - Intermediate (residence), Joint Forces Staff College, Norfolk, Virginia
2004 Air War College (correspondence), Maxwell AFB, AL

ASSIGNMENTS:

1. Nov 1988-Nov 1989, Undergraduate Pilot Training, Columbus AFB, MS.
2. Jan 1990-Mar 1990, Lead-in Fighter Training, Holloman AFB, NV
3. Apr 1990-Sep 1990, Initial F-16 Training, MacDill AFB, FL.
4. Nov 1990-Jan 1995, Scheduler, Assistant Weapons Officer, SOF, F-16 Instructor and Evaluator Pilot, 480th/22nd Fighter Squadron, 52nd Operations Group, Spangdahlem AB, Germany.
5. Jan 1995-Aug 1998, Scheduler, Special Assistant to the OG/CC, Flight Commander, F-16 Instructor/Evaluator/FCF Pilot and SOF, 61st Fighter Squadron, Luke AFB, AZ.
6. Oct 1998-May 2001, Thunderbird #4/Slot, Acting Thunderbird #7/Operations Officer, USAF Aerial Demonstration Squadron, Nellis AFB, NV.
7. Jul 2001-Jun 2002, Student, Class President, Air Command and Staff College, Maxwell AFB, AL.
8. Jul 2002-Mar 2004, Chief of Close Air Support and Air Interdiction, Operations Plans Branch, CAOC 7, Larissa, Greece.
9. Apr 2004-Jun 2004, Student, Joint and Combined Warfighting School-Intermediate, Joint Forces Staff College, Norfolk, VA.
10. Jul 2004-Oct 2004, Requalification F-16 Training, Luke AFB, AZ.
11. Oct 2004-Present, Operations Officer, 52nd Operations Support Squadron, Spangdahlem AB, Germany.

MAJOR AWARDS AND DECORATIONS:

Defense Meritorious Service Medal

Meritorious Service Medal with one oak leave cluster
Air Medal
Aerial Achievement Medal
Air Force Commendation Medal
2002 Secretary of the Air Force Leadership Award - Air Command and Staff College

EFFECTIVE DATES OF PROMOTION:
Second Lieutenant-11 Aug 1988
First Lieutenant-11 Aug 1990
Captain -11 Aug 1992
Major-1 Aug 1998
Lieutenant Colonel-01 Feb 2003
(Current as of 20 Jun 05)

JEFFREY D. LOWERY

Lt Col USAF A-10 Demonstration Team Leader/Pilot

Jeffrey Lowery enrolled at Clemson in 1979 and was graduated in 1983 with a BS Math degree. He was a member of the AFROTC and while in the AFROTC program he completed the AFROTC Flight Instruction Program (FIP) conducted there at the Oconee-Clemson airport. He was not a member of the flying club but after receiving his 2nd Lt commission he was assigned to active duty for pilot training. His military service and flying career highlights are as follows:
1983-1984 Undergraduate Pilot Training (UPT)
Vance AFB, OK T-37 and T-38
1984 Pilot Instructor Training (PIT) Randolph AFB, TX T-38
1984-1988 25th Flying Training Squadron Vance AFB, OK T-38
1988 Lead-in Fighter Training (LIFT) Holloman AFB, NM AT-38
1988 A-10 Flying Training Unit (FTU)
Davis-Monthan AFB, AZ A-10
1989-1991 92nd Fighter Squadron (Skulls) Bentwaters AB, UK A-10
1991-1994 75th Fighter Squadron (Flying Tiger Sharks) Pope AFB, NC A/OA-10
1997-1999 358th Flying Training Squadron (Lobos)
Davis-Monthan AFB, AZ A/OA-10
1999-2000 12th AF A-10 Demonstration Team Leader/Pilot
Davis-Monthan AFB, AZ A-10

Jeff has over 1500 hours in the T-38 and over 1800 hours in the A-10. He also flew 28 combat sorties in Operation Provide Comfort (1991). Also, while flying as the Demo Pilot (1999-2000) he flew as part of Air Combat Command's (ACC) Heritage Program which included formation flying with vintage aircraft like the P-51 Mustang, P-38 Lightning, and F4U Corsair. The pilots of these vintage aircraft included legends like Gen Bill Anders and Frank Borman former Apollo 8 astronauts.
Jeff is now assigned in the Pentagon's Modeling and Simulation Policy Division SAF/XCOM Warfighter Rediness Branch of the Pentagon.

LT COL BRADLEY (BONES) McCOY

Bradley, son of Lt Col Ron McCoy USAFR-Ret, graduated from Clemson University in 1987 with a degree in Computer Information Systems and was active in AFROTC at Clemson all four years. After receiving his commission as a 2nd Lt he entered the Flight

Screening Program at Embry-Riddle Aeronautical University, Daytona Beach, FL. He then was assigned to Pilot Training Class 89-06 from March 8 till March 89 at Vance AFB, OK.

Initally he was assigned to the F-16 then he was assigned to the F-15 where he participated in Fighter Training at Holloman, summer of 1989 as Red Devils. Then he was assigned to F-15 RTU: 426 Tactical Fighter Training Squadron, Luke Air Force Base (Killer Claws). His next assignment was to Langley AFB's 71st Tactical Fighter Squadron (Ironmen), 1989 - 1993. This was the First Squadron to deploy to DESERT SHIELD / STORM 4 months after he became Mission Ready. He was number 2 in a 6 aircraft deployment and was the second person to land in the theater.- He ended up with nearly 300 hours of combat flying time (no kills, but several close calls).

He participated in SOUTHERN WATCH and checked out as a Mission Commander and Instructor Pilot.

Next he was assigned to the 95th Fighter Squadron (Boneheads), 1994-1996. Here he was Academic Instructor, Flying Instructor, and eventually Flight Commander.

His next assignment was to the USAF Weapons School at Nellis AFB, Nevada (Jan 96 thru July 96).

His next assignment was to the 12th Fighter Squadron (Dirty Dozen), July 96 thru Jul 1999. He was Squadron Weapons Officer for 2.5 years.

During this assignment the Squadron was undefeated in Turkey Shoots for 18 straight events during this time.

He participated again in Iraq in OPERATION NORTHERN WATCH. He finished the time there as Assistant Director of Operations. He was promoted from Captain to Major one year below the Zone.

He was then assigned to the USAF University at Maxwell, AFB, Montgomery, Al. to attend the Air Command and Staff College, 1999-2000 and was graduated.

He was next assigned to Peterson AFB to work with NORAD (Sept 2000- Sep 2002) where he was Chief, Special Technical Operations and later Chief, Air Sovereignty Operations. He was there before, during, and after 9/11. While there his boss was a fellow Clemson grad, Col Mike Cook. Together they created what is now the NORAD Operations Center.

After this assignment he went to 433rd Weapons Squadron (Satan's Angels) Sep 2002 - Jul 2005, where he became the Commander in March 2003 and remained there until June 2005.

He then was assigned again to the Air University to attend the Air War College and was graduated in July 2006.

His next assignment was to Langley AFB's, the 5th Generation Fighter Division in July 2006.

Here he was initially Deputy Chief, 5th Generation Fighter Division and is now Chief, F-22 System Management Organization - They develop the requirements for the F-22.

He has some 2000 hours in the F-15 series aircraft (A,B,C, and D models)

Decorations:

Defense Meritorious Service Medal
Meritorious Medal with Oak Leaf Cluster
Air Medal with 3 Oak Leaf Clusters
Aerial Achievement Medal with 3 Oak Leaf Clusters
Air Force Commendation Medal with Oak Leaf Cluster
AF Outstanding Unit Award with 3 Oak Leaf Clusters
Combat Readiness Medal with 3 Oak Leaf Clusters
National Defense Service Medal with Oak Leaf Cluster

Armed Forces Expeditionary Medal
Southwest Asia Service Medal with 2 Oak Leaf Clusters
Global War on Terrorism Service Medal
Air Force Overseas Ribbon Short
Air Force Overseas Ribbon Long
AF Longevity Service Medal
Small Arms Expert Marksmanship Ribbon (Pistol)
AF Training Ribbon
Kuwait Liberation Medal Kingdom of Saudi Arabia
Kuwait Liberation Medal Government of Kuwait

DR. MIKE SMITH, PhD

Aeronautical Researcher and Aviation Business Executive

Dr Smith is a senior aviation executive with over 42 years of experience in aircraft development and certification and has been a pioneer in the field of composite material processes development and certification methods for General Aviation(GA) and special use aircraft. He has been involved in six light plane aircraft development and certification programs and has performed extensive research and development for NASA, DOD, FAA, USDA, and the U.S. Aircraft industry. He worked for the Raspet Flight Research Center at Mississsippi State University for 8 years and operated a consulting engineering business for twenty years.

Dr. Smith, the founder of Global Aircraft, has served as President of Global Aircraft Corporation since 1993 where he managed the development of an all composite GA airplane and developed a revolutionary composite propeller for GA airplanes that is more efficient and quieter than typical metal propellers. He has been a key participant in the formation of the NASA AGATE program to revitalize General Aviation and has served as the Leader of the NASA/FAA Task Group to develop certification guidlines for GA Composite Airplanes.

He has provided consulting services as a Professional Engineer since 1965. He was the Principal Investigator or Project Manager on 19 AGATE projects since 1994 and the Principal Investigator on 11 NASA Phase 1 SBIR projects. 3 NASA Phase II projects, 2 NASA STTR Phase I, and 2 NASA Phase II since 1995. He has also been PI on a NSF SBIR Phase I Project and a DOD Phase I SBIR project. Two of the NASA SBIR projects have led to commercial products that are currently in production by Global. One of the NASA STTR Phase II projects has led to commercialization and another is in further Phase III development prior to production.

Dr Smith has published over 55 technical reports and articles on various aspects of aircraft design, development, and flight testing, design and development of composite propellers, and the development of composite processes. Several of the listed NASA projects have related to aircraft engine testing and development of more efficient engine exhaust systems and mufflers. One NASA project involved the testing of a novel diesel rotary engine and the design of a 250 BP rotary engine for high performance GA airplanes and UAV's.

Dr. Smith has served as President of Advance Technologies Inc since 2001 and its predecessor, Advanced Developments Inc from 1966 to 1993. ADI and ATI have engaged in a broad range of consulting engineering projects including aircraft design and development, composite process development and production of a low cost Head-Up Display (HUD) for GA aircraft that will reduce spatial disorientation in IFR and in marginal VFR flight operations.

This work has been supported by NASA Aviation Safety Program and the NASA

Small Airplane Transportation program, in addition to internal R&D funding.

ATI has recently been engaged in the development of alternative energy production using agriculture biomass wastes to generate electrical power. This work includes a novel rotary engine expander to power an electric power generator and development of methods to accelerate digestion of biomass wastes to increase the efficiency of waste recovery processes. Three patent applications have been submitted on the axial-vane rotary technology.

From 1972 to 1993, Dr. Smith was an Assistant Research Engineer with the Raspet Flight Laboratory at Mississippi State University where he was engaged in aviation, aeronautical, aerospace, and composite materials research. From 1963 to 1968, Dr. Smith lead a research project to evaluate the aerodynamics and flight stability of all models of Agricultural aircraft manufactured in the United States. Several improvements in the safety and utility of agricultural airplanes resulted from the project.

He also participated in the development of composite processes related to the development of the XV-11A STOL airplane and was a leader in the investigation and optimization of the suction boundary layer control system on the XV-11A. Dr. Smith lead the investigation of 3-Dimesional Boundary Layer Flow over Yawed Wings and the Application of swept wing theory to boundary layer flow over helicopter blades.

From 1967 to 1972, Dr. Smith was the leader of a project to determine the dynamic geometry of exhaust clouds from static firings of Apollo S-I and S-II rocket engines and the development of computer models to predict the dynamic geometry of such exhaust clouds under arbitrary meteorological conditions.

Dr Smith is a member of the American Institute of Aeronautics and Astronautics, the Society of Automotive Engineers, the National Society of Professional Engineers, the Mississippi Society of Professional Engineers, and is a registered professional engineer in the State of Mississippi. He has served on numerous technical committees in AIAA and NASA programs and has served as a consultant to the U.S. and Canadian governments regarding aircraft safety and operations. He is still an active pilot and has owned a number of aircraft over the past 40 years, these being a Piper Cherokee 235, a Piper Comanche 259, a Piper Seneca, a Glasair II 180, and the Polish ISKRA Jet aircraft. He has logged about 6100 hours and has flown the SISSU High Performance Sailplane, numerous experimental aircraft as Whitman Tailwind, Turner T-40, RV-6, RV-6a, Glasair, Glasair II, Glasair III, the XV-IIA, Global GT-3. In addition he has flown twins such as the DC-3, Twin Beech, Beech 250, and 350, the Aero Commander 680. During 1964-1969 he flight tested the current Agricultural aircraft as the PA-150, Pawnee, Cessna Ag-Wagon, Call Air A-9 and the B-2, the Snow S-2, the Gruman AG-Cat, and the Stearman 450.

This man has been very active in aviation research, development, and manufacturing of unique aircraft and related products for aviation and Industry over a number of years and the Clemson Aero Flying Club experience appears to have had quite an influence in his career.

MILES HUNTER WATKINS
Lt Col USAFR-Ret

Miles grew up in Pendleton, SC, son of Thomas Green and Bessie Hunter Watkins. He was very interested in aviation as a child and spent a lot of his youth observing the Eastern Airlines DC-3's flying along Airway 6 Green, the Atlanta, Ga. to Charlotte, NC airway and telling people that he wanted to be a pilot. He was graduated from Pendleton High School in 1938 and enrolled in Clemson to study for a degree in Mechanical Engineering. While attending Clemson the Civilian Pilot Training Program (CPTP) Miles was introduced by the Civil Aeronautics Adminstration, the U.S. Department of Commerce in conjunction with the Reserve Officers Training Program. He immediately signed up, passed his physical, and began his training to be a pilot. He completed all required training and obtained his private license in September 1940. He was invited and became a member of the Clemson Flying Cadets. He next took the Secondary Course for Advanced Acrobatic and Cross Country flying instruction and received his Certificate in June 1941. Both courses were under the direction of Dean S. B. Earl's Mechanical Engineering Department and Southern Airways, Inc of Anderson, SC.

He was commissioned 2nd Lt. on May 23, 1942 and reported for duty at Hunter Field, Savannah, Ga. From there he went to Army Air Force pilot training and graduated from Pilot Class 43-D, Turner Field, Albany, Ga., receiving his wings on 29 April, 1943. He remained at Turner Field as an Instructor for the Curtis AT-9 and the Beechcraft AT-10. He later instructed in the North American B-25 at Turner Field until 1943. He transferred to Hqs. Eastern Flying Training Command as Staff Instructor Pilot (B-25) and Staff Pilot for Airlift of EFTC personnel. This duty allowed him to continue his pilot proficiency in a number of aircraft as the AT-6's, AT-7's, AT-11's, a C-45, a C-47, a C-53, a C-60, a P-40, a P-51, a B-25, and a B-17.

He was selected as Aide de Camp for BG William S. Gravely, Commander of the Eastern Flying Training Command. He was promoted to Lt and then Captain and then assigned to Hqs, 1503rd Air Transport Wing of the Air Transport Command (ATC), Harmon Field, Guam in June 1946 and then moved with the Wing to Kadena Air Force Base, Tokyo, Japan, in November, 1946. Here he was assigned to the Air Traffic Division and he was responsible for coordinating the requirements of Air Passage and Air Freight Sections to all Air Bases in the Wings (Harmon, Haneda, Naha, Clark, and Shanghai. In early 1947 the around the world Embassy Flight was established with units in Saigon, Vietnam; Bangkok, Thailand; Calcutta, New Deli and Karachi, India; and Dhahran, Saudi Arabia. Later in the year the Embassy Flight was extended to Asmera, Eritrea. In October, 1948 Miles became the Commander of the ATC Detachment at Naha, Okinawa. This was the main route for personnel returning from China and high ranking military and civilian people.

He transferred in July 1949 to the 1601 Air Transport Group, Continental Division, Military Air Transport Service (MATS), Brookley Air Force Base, Mobile, Al. He attended the Air Tactical School at Tyndall AFB, Fl from October thru December 1949 and was promoted to Major in December 1951. He then was transferred to the 1291st Air Transport Sq (C-124's) Continental Division, MATS, McCord AFB, WA. He later became the Operations Officer of the 32nd Air Transport Squadron (C-124) in May,1953 he was transferred to the 22nd Troop Carrier Sq.(C-124), 374th Troop Carrier Wing, Tachikawa AFB, Japan. While at Tachikawa in the TWC Operations Division he became the Assistant Operations Officer and briefed the Commanding General, TCD and Staff. In 1955 he transferred to Headquarters, Pacific Air Forces, Hickam, AFB, Hawaii where he was assigned to the Plans and Program Office.

In June 1957 he was transferred to the Air Command and Staff College, Maxwell AFB, Montgomery, AL where he completed the ACSC course and was next assigned to the Pentagon, Headquarters United States Air Force where he remained for 4 years. He was promoted to Lt Col in June 1960 and in July 1962 he transferred to the 1st Military Airlift Sq, 1607th Military Airlift Wing (MAC), Dover AFB, DE. There he attended the C-133 Heavy Transport Training Unit School and assumed Command of the 1st Military Airlift Squadron in May 1963. In March 1965 he was transferred to Hq. US Air Forces, Europe (USAFE) and placed in charge of the Air Transport Division. More pilots were needed in Vietnam so he transferred 536th Tactical Airlift Squadron in Vung Tan, Vietnam. While en-route he completed the Advanced Flying Course for the DeHavilland aircraft the C-7A, Caribou. In March he assumed command of the 536th Squadron. On Christmas Day 1968 in coordination with the USO, Miles and five of his aircraft dropped foot lockers full of Christmas goodies to US Army and Navy units in the Mechon Delta that could not be reached by normal transportation because of moonsoonal rains and enemy actions. In January 1969 he was transferred to Kelly AFB, Texas and on September 1, 1969 he retired from the USAF.

After his retirement he began a second career in the financial world of banking, working for the National Bank of Commerce of San Antonio, Texas from 1970 thru 1981 There he assisted in setting up the MasterCard division and he later advanced to Assistant Vice President in the Marketing Division.

He is a life member of the Order of Daedalians, Veterans of Foreign Wars (VFW), Military Officers Association of America (MOAA), member of Air Force Association (AFA), Experimental Aircraft Association (EAA), the Canyon Lake United Methodist Church, and other civic organizations.

His military decorations:

Air Medal with 4 Oak Leaf Clusters
Air Force Commendation Medal
AMERCM
WWII Victory Medal
Army Occupation Medal
Korean Service Medal
United Nations Service Medal
National Defense Service Medal with 1 BSS
Republic of Korea Presidential Unit Citation
SAEMR
AFSLA with 5 Oak Leaf Clusters
VSM with 2 BSS
Republic of Vietnam Campaign Medal.

Total Flying Hours:

Military	7885
Civil	185
Total	8040

Military Combat Flying Hours 506.7

He and his wife Margaret Ramey are the parents of two children, Martha and Hunter Jr. Martha's husband is Bruce Head and Hunter, Jr and his wife Becki have a son Miles Hunter Watkins III.

PROFESSOR JOHN J. HARPER

Professor of Aeronautical Engineering at Georgia Tech

Professor John J. Harper was born and grew up in Seneca, S.C., and became fascinated with model airplanes as a child. He was one of the charter members of the Seneca Model Airplane Club that was formed in July 1939 there in Seneca The club entered 9 model airplanes to compete in the Tri-State Model Airplane contest held in Greenville, SC during August, 1959, John entering a gasoline powered model. He was graduated from Seneca High School in 1936 and began his studies in Mechanical Engineering and graduated with a Bachelor's degree in Mechanical Engineering from Clemson A& M College in 1940 and his Masters Degree in Aeronautical Engineering from Georgia Tech in 1942. He began his tenure as a professor at Georgia Tech in 1946, specializing in aerodynamics and aircraft design. Later he became Professor of Aeronautical Engineering.

Throughout his four decades as a professor at Tech, John taught undergraduate classes on the fundamentals of aerodynamics. He was the Director of Tech's 9 foot diameter wind tunnel that was used for academic and industrial research during the early 1950's. Some of his former students who later became Astronauts were Scot Horowitz, Space Shuttle Pilot, Richard Truly, Space Shuttle Commander STS-2/OSTA 1 Flight and John Young, who was the Commander for the First Space Shuttle Flight STS-1 as well as the Commander of Apollo 16-a mission to land on the Moon and conducted Scientific Experiments while on the Moon. John Harper was a private pilot and built his own plane and was particularly interested in aircraft performance, stability and control.

John spent hours building meticulously detailed wooden ships for museums and he also built radio-controlled airplanes prior to his retirement from Georgia Tech in 1986.

John was a registered professional engineer, a member of the American Institute of Aeronautics and Astronautics, a member of the Experimental Aircraft Association, and an Honorary member of the Subsonic Testing Association.

Professor Harper died in 1998 in Atlanta, Ga.

KEITH PARRISH

Captain USAF-Distinguished Military Aviator and Author

Keith R. Parrish was born in Spartanburg, South Carolina, in 1946. Mr. Parris graduated from Clemson University with a Bachelor of Science Degree in 1969 with Highest Honors and was a Distinguished Military Graduate in AFROTC. He was awarded a Regular Commission and served nearly 8 years as a fighter pilot, flying F-4E Phantom II aircraft, in Europe, Asia and the United States. As an Air Force captain, he earned numerous commendations, including a Meritorious Service Medal and Top Gun Aircraft Commander, Seventeenth Air Force, United States Air Forces in Europe. He married Janet K. Baugh, of Waco, Texas, in 1974 while serving in the Federal Republic of Germany. They lived in Europe and Asia before moving to San Antonio, Texas, in 1977 after Captain Parris resigned his commission and was honorably discharged from the Air Force. He returned to school and graduated at the top of his class from The University of Texas at San Antonio in 1979 with an MBA Degree. He earned his CPA certificate in 1984 and was honored as one of the Top Ten recipients to be certified in the State of Texas. He worked for 21 years in the Gas and Gas Liquids Industry where he served as Director of both financial and operational areas at Valero Energy Corporation and Pacific Gas and Electric Corporation's Texas subsidiaries.

In 1998 he fell seriously ill and underwent two surgeries to remove and correct symptoms associated with a tumor located in the pineal region of the brain. His successful recovery from this challenge had a profound impact on his life. He retired from industry in December 2001 and began devoting himself to his lifelong passion of writing. His first book, TORI And Other Poems And Short Stories, was privately published and distributed. This book, The Servants Of Freedom, is the culmination of more than 3 years of creative endeavor and reflects his admiration and appreciation for the sacrifices of the men and women of our Armed Forces. In this novel of 1970s Europe, he uses simple, yet powerful language to make timeless themes of adversity and courage live in the minds of his readers. His writing style reflects his admiration for the literary works of such outstanding 19th- and 20th-century novelists and poets as Ernest Hemingway, Eudora Welty, William Faulkner, Robert Frost, Emily Dickinson and William Wordsworth.

OTHA H. VAUGHAN JR.

NASA Aerospace Engineer and Atmospheric and Space Science Researcher Lt Col. USAFR-Ret

Mr. Vaughan retired during 1999 after working in the Space Program for 39 years with NASA Marshall Space Flight Center in Huntsville, Alabama. He had previously worked for the USAF Air Proving Ground Command, Eglin AFB, Florida and the Army Ballistic Missile Agency at Redstone Arsenal in Huntsville, Alabama prior to transfering the NASA. During his tenure at Redstone he was a member of the von Braun Research and Development Team that developed the Redstone, Jupiter C, Jupiter, Juno, and the Saturn IB missiles. While employed by the Army he worked in the areas of Aerodynamic Heating Studies, Test Flight Evaluation of Missiles, and in specialized Cooling System designs for ballistic missiles.

He graduated from Clemson A & M College in June 1951 with a degree in Mechanical Engineering. While at Clemson he was an AFROTC Distinguished military student and upon graduation was commissioned in the reserves as a 2nd Lt USAFR. He entered active duty as soon as he was graduated and served as an Aircraft Maintenance Officer for the 25th Weather Squadron at Warner Robins AFB, Robin, Ga. He entered USAF Pilot Training Class 52G but could not complete his pilot training due to a medical condition and was released from training and was then assigned to a Jet Engine Engineering Research and Development Group at Tinker AFB, Oklahoma. This Air Force Base was under the Air Material Command and specialized in the overhaul and repair of aircraft and jet engines for the USAF. Mr. Vaughan remained at Tinker till March 1953 and was released from active duty to attend graduate, however remaining in the reserves. He attended Clemson for a few years working on a Masters Degree in Mechanical and after completing all of his required studies except his Thesis he went to work for Vitro Corporation at Eglin AFB, Florida as a Research and Development Test Engineer, developing specialized weapon systems to be delivered by aircraft. He later worked in the Air Proving Command there at Eglin before transferring to the Army Ballistic Missile Agency at Redstone Arsenal in Huntsville, Al. While employed at ABMA he was released to return to graduate school for a short time to complete his Thesis and was awarded his Masters Degree in Mechanical in January 1959 from Clemson A & M College, Clemson, S.C.

When NASA establised the NASA Marshall Space Flight Center(MSFC) in Huntsville, Al. he became a charter member in June 1960 by transfering from ABMA with the von Braun Research and Development Team to NASA. While at ABMA and NASA MSFC he continued his aerodynamic heating and flight evaluation studies and became a Supervisory Aerospace Engineer directing the work of other engineers under his supervision. Later he became involved in the Apollo program in the development of design criteria for space vehicles and for lunar and planetary surface exploration vehicles. He was a member of the Lunar Scientific Advisory Team for the first Cis-Lunar Apollo 8 mission that circled the Moon and returned the astronauts back to the Earth. He developed design criteria for the Lunar Rover that was used for the Scientific Exploration of the Lunar Surface during the Apollo Missions Apollo 15, 16, and 17. He provided design criteria for the construction of a Lunar Driving Simulator that was used to train the astronauts to drive the Lunar Rover. In addition to these contributions he developed a number of scientific demonstration that were accomplished during the Skylab Program. To accomplish the feasibility of these experiments he participated as a test subject flying in the NASA KC-135 Zero Gravity Aircraft to conduct these demonstrations and then develop the procedures for accomplishing these demonstrations. He served as Mission Support Scientist for the Shuttle Space Lab 3, and Assistant Mission Scientist for the Shuttle Space Lab J, a joint Japaneese National Space and Development Agency and NASA Space

Shuttle mission. Through his government service he has received numerous NASA Group Achievement Awards, and Special Service Awards. He participated as a member of the Apollo Debriefing Teams and was a member of the Challenger Space Shuttle STS-51L Accident Investigation Team.

Vaughan was a former member of the American Rocket Society, joining in 1957 and later becoming a member of the American Institute of Aeronautics(AIAA) in 1957. He was inducted as an Associate Fellow of the AIAA in 1990 and was selected by his peers to be the AIAA Aerospace Engineer of the Year for 1996. He was later awarded for the 1998-1999 Year the Hermann Obert Award(a Certificate of Acheivment and a bronze bust of Dr. Obert---known as the Father of Rockerty in Germany in the 30's and a mentor to Dr. von Braun). This award was given Mr. Vaughan for his outstanding Research in Atmospheric Electricity and the discovery of the Red Sprites and the Blue Jets that were seen to occur above Severe Thunderstorms by using the onboard TV cameras of the Space Shuttle while conducting his research program, the Mesoscale Lightning Observations Experiment (MLE). In February 2001 he was inducted into Clemson University's prestigious Thomas Green Clemson Academy of Engineers and Scientists. In January 2006 he was the recipient of the AIAA's Losey Atmospheric Sciences Award. He has authored or co-authored 90 technical reports and/or published articles on aerodynamic heating studies, missile flight test evaluations, lunar exploration, lunar mobility studies, zero gravity fluids research, atmospheric cloud physics experiments and more recently in Atmospheric Electricity research from ground based and space based observations. He is a General Aviation Pilot and has over 600 hours and is a member of a number of professional societies such as Sons of the American Revolution, Life Member of the Air Force Association, a Charter member of the National Aviation Hall of Fame in Dayton, Ohio, Life member of the Antique Aircraft Association, Life member of the Reserve officers Association, a Lt Col USAFR-Ret, a Lt Col in Civil Air Patrol, Member of the Experimental Aircraft Association, the Dragon Lady Association, a Shriner and Mason. Vaughan was a consultant for two TV productions, *Mega Lightning* and *Moon Machines*. He is Methodist and is active in his church. He is married to the former Betty Frances McCoy of Atmore. Al. and is the father of three daughters and four grand children (three grandsons and a granddaughter).

DR. JAMES F. SUTTON, PhD

Mechanical Engineer, Educator, Inventor

Dr James F Sutton graduated from Clemson, Class 1945 with a BS in Mechanical Engineering with a minor in Electrical Engineering and his Officers Commission thru Army ROTC and then entered military service and served in the Pacific Theater of Operations. After his discharge he returned to Clemson to teach in the Mechanical Engineering School. He continued to teach and later received his Master's Degree and PHD from the University of Michigan. He left Clemson in 1955 and worked at the Lockheed Aircraft Georgia Company. He was awarded several patents, including the one for the propulsion system for the Hummingbird, a vertical takeoff aircraft (VTOL) that began its development by Lockheed as a private company and Lockheed later received a development contract in 1961 from the US Army to build and fly two test prototypes. The original purpose of the Hummingbird XV-4A was to replace the helicopter that was then being used in Vietnam. Two aircraft were designed and fabricated for test flights, and were ready for testing by November of 1962. During May 23, 1963 the first free hover test flight was accomplished. However during a later test flight on June 10, 1964 an aircraft was destroyed, killing the pilot, and the Army canceled the project. The first Hummingbird concept as designed never got into production. Dr.Sutton later was the first Director of Research at Lockheed and later he was heavily involved in the C-5 project. He left

Lockheed in 1971 to return to academia to become the Dean of Engineering at University of Wyoming. During the three years while he was at Wyoming he began the development of a Transit concept called Transette. He moved back to Georgia working at the Georgia Tech Research facilities to continue his Transette research being funded by a number of federal research grants.

He died unexpectedly on January 28,1985.

DR. WALTER F. JONES, PhD

Dr. Walter F. Jones, a member of the Senior Executive Service, is Director, Plans and Programs, Air Force Research Laboratory, Wright-Patterson Air Force Base, Ohio. Dr. Jones is responsible for developing and managing the processes that define AFRL's $3 billion annual investment in technologies for future Air Force systems. These systems include space, weapons, aeronautics, and command, control, communications, computers, intelligence, surveillance and reconnaissance.

Dr. Jones has held a wide variety of positions in government and academia. He most recently served as Director, Aerospace and Materials Sciences, Air Force Office of Scientific Research, Arlington, Va. In this capacity, Dr. Jones planned, coordinated and executed a $55 million basic research program, including solid mechanics, fluid mechanics, materials science and propulsion. He has also served as Senior Program Analyst with the Office of the Deputy Director of Central Intelligence for Community Management. He has held several positions with the Air Force, including Deputy for Research Sciences with the Office of the Assistant Secretary of the Air Force for Acquisition, and Deputy for Science and Technology with the Office of the National Security Space Architect. In addition, Dr. Jones has held faculty positions at the University of Florida and the University of Tennessee.

EDUCATION

1978 Bachelor's degree in mechanical engineering, Clemson University, Clemson, S.C.

1981 Master's degree in engineering mechanics, Clemson University, Clemson, S.C.

1982 Doctor of Philosophy degree in engineering mechanics, Clemson University Clemson, S.C.

1998 Leadership for a Democratic Society, Federal Executive Institute, Charlottesville, Va.

1999 Master's degree in national resource strategy, Industrial College of the Armed Forces, Fort Lesley J. McNair, Washington, D.C.

1999 Senior Acquisition Course, Industrial College of the Armed Forces, Fort Lesley J. McNair, Washington, D.C.

2005 Certificate in Legislative Studies, Government Affairs Institute, Georgetown University, Washington, D.C.

CAREER CHRONOLOGY

1. August 1982 - August 1983, visiting assistant professor, Department of Mechanical Engineering, Clemson University, Clemson, S.C.
2. August 1983 - August 1986, assistant professor, Department of Engineering Science and Mechanics, University of Tennessee, Knoxville
3. August 1986 - August 1988, associate professor, Department of Aerospace Engineering, Mechanics and Engineering Science, University of Florida, Gainesville
4. August 1988 - July 1991, development staff member, Engineering Technology Division, Oak Ridge National Laboratory, Oak Ridge, Tenn.
5. July 1991 - October 1995, program manager, Directorate of Aerospace and Materials

Sciences, Air Force Office of Scientific Research, Bolling Air Force Base, Washington, D.C.

6. October 1995 - August 1998, Deputy for Research Sciences, Office of the Assistant Secretary of the Air Force for Acquisition, Arlington, Va.

7. August 1998 - June 1999, student, Industrial College of the Armed Forces, Fort Lesley J. McNair, D.C.

8. June 1999 - June 2001, Deputy for Science and Technology, Office of the National Security Space Architect, Office of the Assistant Secretary of Defense for Command, Control, Communications and Intelligence, Chantilly, Va.

9. June 2001- January 2002, Senior Program Analyst, Program Assessment and Evaluation Office, Office of the Deputy Director of Central Intelligence for Community Management, Washington, D.C.

10. January 2002 - August 2005, Director, Aerospace and Materials Sciences, Air Force Office of Scientific Research, Arlington, Va.

11. August 2005 - present, Director, Plans and Programs, Air Force Research Laboratory, Wright-Patterson AFB, Ohio

PROFESSIONAL MEMBERSHIPS AND ASSOCIATIONS

Armed Forces Communications and Electronics Association International
American Academy of Mechanics
Associate fellow, American Institute of Aeronautics and Astronautics
Fellow, American Society of Mechanical Engineers
Senior Executives Association
Executive board, Society for Experimental Mechanics

(Current as of January 2006)

DR. FORREST J. AGEE, PhD

Retired Sep. 1, 2006

Dr. Forrest J. "Jack" Agee, a member of the Senior Executive Service, is Director of Physics and Electronics, Air Force Office of Scientific Research, Arlington, Va. He is responsible for the $80 million Air Force basic research program in physics and electronics, assuring the excellence and relevance of a broad research portfolio. His program encompasses hundreds of university grants and supports work of undergraduates and more than 300 graduate students. This work also supports basic research within the Air Force Research Laboratory, industry and overseas. The research includes novel space-craft engineering, semiconductor device research, nanotechnology, electronic sensors, polarimetry research, gravitometry, lasers, plasmas, high-power radio frequency sources, high-temperature superconductivity engineering, atomic and nuclear physics, space optics, and imaging and opto-electronics. Agee plans, coordinates and executes a research program conducted by scientists in academia, industry and Air Force laboratories. He was appointed to the SES in 1998.

Prior to beginning his career with the Air Force, Dr. Agee worked as a U.S. Navy physicist in acoustics related to silencing submarines. He also worked in low-temperature physics for the U.S. Army at Fort Belvoir, Va. In the 1970s, he joined the Harry Diamond Laboratory in Adelphi, Md. He handled many assignments of increasing scope and responsibility, initially in nuclear electromagnetic pulse hardening and testing for the Safeguard Anti-Ballistic Missile System and for Army tactical systems. As the Technical Director, he led a major, strategic, Defense Nuclear Agency Electromagnetic Pulse Test for the Commander in Chief of Pacific Forces. For three years he managed strategic command, control and communications programs at The BDM Corp., including the Airborne

Command Post. In 1982 Dr. Agee became Director of the Aurora Radiation Test Facility that tested the Peacekeeper missile and other systems at the Harry Diamond Laboratory. In the Army's high-power microwaves program, Dr. Agee led research and development efforts in sources, and in 1990, became Director of the Army program. His efforts led to developing the AN/VLQ-9 and AN/VLQ-10 Shortstop electronic warfare systems during operations Desert Shield and Desert Storm. The Shortstop System protected U.S. forces in Bosnia and now has seven versions, including two that are deployed in Iraq in a counter-IED role. Dr. Agee formed the Joint Directors of Laboratories Panel on Directed-Energy Weapons in 1990, and chaired the panel until 1993. That same year, he began his Air Force career in the Directed-Energy Program at Phillips Laboratory at Kirtland Air Force Base, N.M., where he led the research program in high-power microwave sources.

Dr. Agee's principal fields of interest are in electrical engineering, pulsed power, microwave generation, electromagnetics and superconductivity. In physics, his interests include nuclear physics, plasma physics, semiconductor physics and lasers. He has written or co-written more than 200 publications and inventions in these areas.

EDUCATION

1963 Bachelor of Science in physics, high honors, Clemson University, Clemson, S.C.
1965 Master of Science degree in physics, University of Virginia, Charlottesville
1967 Doctor of Philosophy degree in physics, University of Virginia, Charlottesville

CAREER CHRONOLOGY

1. 1963-1965, physicist in acoustics, U.S. Navy Engineering Experiment Station, Annapolis, Md.
2. 1965-1965, mathematics instructor, Howard Payne College, Brownwood, Texas
3. 1965-1971, physicist in cryogenics, U.S. Army Engineer Research and Development Laboratories, Fort Belvoir, Va.
4. 1971-1979, physicist and electronics engineer, Harry Diamond Laboratories, Adelphi, Md.
5. 1979 - 1982, Manager, The BDM Corp., McLean, Va.
6. 1982 - 1990, Director, Aurora Facility, Harry Diamond Laboratories, Adelphi, Md.
7. 1990 - 1993, Director, High Power Microwave Technical Management Office, Harry Diamond Laboratories, Adelphi, Md.
8. July 1993 - December 1993, Technical Director, Division Chief, Electromagnetic Sources Division, Phillips Laboratory, Kirtland AFB, N.M.
9. December 1993 - October 1995, Chief, Electromagnetic Sources Division, Phillips Laboratory, Kirtland AFB, N.M.
10. October 1995 - February 1998, Technical Director, High Energy Sources Division, Air Force Research Laboratory, Kirtland AFB, N.M.
11. February 1998 - present, Director of Physics and Electronics, Air Force Office of Scientific Research, Arlington, Va.

AWARDS AND HONORS

1963 Phi Kappa Phi Honor Society
1967 Sigma Xi, The Scientific Research Society
1967 Presidents and Visitors Research Prize, University of Virginia
1975 Army Research and Development Achievement Award
1985 Army Meritorious Civilian Service Medal
1987 Strategic Defense Initiative Laboratory Technical Achievement Award, Strategic Defense Initiative Organization
1988 U.S. Senate Certificate of Recognition, Productivity Award Program

1989 Secretary of Defense Productivity Excellence Award
1994 Joint Logistics Commanders Certificate of Merit
2000 Fellow, Institute of Electrical and Electronics Engineers
2005 Life Achievement Award, Association of Old Crows
2006 Outstanding Civilian Career Service Award

OTHER ACHIEVEMENTS
2000 Speaker, "An Innovative Approach to Satellite Technology," NATO SET Symposium
2000 Speaker, Plenary page, "Review of Induced Gamma Emission Research and Recent Results - Implications for Further Research," Proceedings, International Conference on Lasers
2001 Joint U.S. patent, Power Enhancer for Solid State Switched Ultrawideband Array Transmitters
2001 U.S. patent, Ultra-Wide Band RF-Enhanced Chemotherapy for Cancer Treatment

PROFESSIONAL MEMBERSHIPS AND ASSOCIATIONS
American Physical Society
Association of Old Crows
Fellow, IEEE
Former member, Advisory Board to the Dean of Sciences, Clemson University
International Society for Optical Engineering
Former chairman, Joint Directors of Laboratories Technology Panel on Directed Energy Weapons
Former senior U.S. representative, NATO Research Study Group 1 on Military Applications of High-Power Microwaves
U.S. member, Secretariat, International Commission on Induced Gamma Emission
Former chair, Membership Committee, Plasma Sciences Applications Committee, IEEE
Former Elected member of Ex-Com, IEEE Plasma Sciences Applications Committee
Former Chair, Office of the Secretary of Defense Strategic Planning Group for Basic Research in Physics
Former Member at large, Board of Directors, American Institute of Aeronautics and Astronautics

PUBLICATIONS
1997 "Silicon Diodes in Avalanche Pulse Sharpening Applications," IEEE Transactions on Plasma
1998 "Ultra-Wideband Transmitter Research," IEEE Transactions on Plasma
1998 "Evolution of Pulse Shortening Research in Narrow Band, High-Power Microwave Sources," IEEE Transactions on Plasma Science
1998 "High-Power, Repetitive-Stacked Blumlein Pulsers Commutated by a Single Switching Element," IEEE Transactions on Plasma Science
2000 "Method for Detection of X-ray Triggered Decay of the 178m2 Hf Isomer Using a Multidetector Array," Proceedings, 9th Annual International Laser Physics Workshop
2001 Chapter, "HPM Sources: The DOD Perspective," High-Power Microwave Sources and Technologies, IEEE Press
(Current as of August 2006)

HUGHLEY (HUGH) WOODLE, PhD

Chemical Engineering and AOPA
and Experimental Aircraft Association Member

Hugh Woodle became interested in aviation at an early age, about 5 years old and then later building, and flying model airplanes as a child with his next door neighbor Henry Peebles, who later became a pilot in the Army Air Corps flying the Lockheed P-38. Henry was Hugh's mentor before Henry entered in service. Hugh didn't get into aviation till later in his life.

He began his College education at Clemson A& M College in 1947 and graduated with a degree in Chemical Engineering and received his officers commission from Army ROTC as a 2nd Lt. in 1951. While in the ROTC program at Clemson he held a number of responsible positions finally becoming a Lt Col and the Commander of 1st Battalion, Clemson Corp of Cadets. He also was a member of Blue Key, Scabbard and Blade, Senior Platoon, Rifle Team, "The Tiger" Staff and Minor "C" Club.

He also was active in other student functions such as Alpha Chi Sigma, S.A.S.ME, A.I.Ch.E, as well as other College activities also serving as Vice President of the Class of 1951.

His first job after graduation was with Du Pont in 1951 and shortly thereafter he reported to active duty, serving till 1953 with the Army Corp of Engineers. After his military service, he went back to DuPont. After a year he resigned to do graduate work, receiving his Masters and later his PhD in Chemical Engineering there at Virginia Tech. He married his advisor's secretary, the former Sue Spence, and in 1957 he went back to DuPont with a PhD in Chemical Engineering, and began a short career in cellophane production in Richmond, VA.

In 1962 Hugh and four others (two already had their pilots license) formed a flying club and they bought a 1955 Tri Pacer in which he received most of his instruction, receiving his Private Pilots License in 1963. Hugh enjoyed several flying trips back to South Carolina, with his young son as his passenger. Hugh transferred to Textile Fibers Dept and moved to Martinsville, VA, where he then bought into a Cessna 175 in a partnership, and later bought the other half when his partner was transferred.

He kept this aircraft until he was transferred to Chattanooga, and made numerous cross-country trips in this aircraft to visit family in Alabama and South Carolina. With the engine turning at 3000 rpm and a reduction gear swinging that long propeller, the pilot had a good, solid feeling moving through the air at 135-140 mph.

He transferred back to Richmond in 1967, and was very fortunate to find 41 acres of land west of Midlothian, Va. with a 1900' grass strip. He built a new house on the property in 1968, and based the 175 at the grass strip which was designated as VA80. However, he was so busy at the plant there was no time to fly, so he sold the 175. Later, when his daughter married and started a family, he set aside two acres and built a house for them.

The business world calmed down a little, and in 1970 he decided he needed an airplane. Hugh searched Trade-A-Plane for a Piper Cruiser and noticed an ad for a 1946 Cessna 120 based in Raleigh, NC. Learning to fly a tail wheel aircraft took seven hours of instruction, and he was off to the races. N1740N truly provided a re-birth for his aviation adventures and his overall outlook on life.

In 1968 he visited Jim Bede in Ohio to look at his BD4 homebuilt and decided he could adapt Bede's construction techniques to design and build the airplane that he really wanted. Hugh had always liked the concept of a push-pull design like a Cessna 337. Hugh then joined EAA Richmond Chap 231 and started designing a low wing, push pull homebuilt with gross design weight in the range of 2500-2600 lbs. A regional FAA FSDO was based at Richmond airport, and he now would have direct access to the folks who would have to approve his work. He documented 10,000+ hours of design and construction over five to six years, to the point that he was ready to order engines, when the insurance companies told him that since it was a one of a kind, the insurance premiums would have to be greater than he would be willing to pay, even though he had log books for all the parts signed off by FAA inspector. He salvaged what generic parts he could, and sent the rest to a scrap yard. He began to study Stress and Strain again along with basic aeronautical design principles, and he learned a lot about why an airplane can fly and stay in the air.

In 1983 the annual for his Cessna 120, 40N turned up a large number of AD notes that he didn't have time to fool with, so he put it up on blocks and bought a Cessna 150 to fly. In 1985 he bought another Cessna 175 and sold the 150. In 1987 he took a big step and decided to completely rebuild the 120 and make it his "traveling" airplane. This effort consumed most of his spare time, and then when he retired in 1990 the work consumed all of his time.

After many hours of hard, frustrating but enjoyable work, he finally got the final approval and signoff in January, 1992. The aircraft had been completely dismantled, every rivet and inch of surface checked with a magnifying glass, wings rebuilt and recovered, the Continental engine 85 replaced with a Continental 0200A, and a very unique full IFR panel installed. Later he even installed vortex generators on the wings, thus reducing the stall speed down to 35-40 knots.

40N became the focus of his aviation interest. Hugh built a 40' x 45' fully enclosed hangar/shop six feet from the house. It took a little time for him to appreciate how lucky he was to have such an arrangement. His wife flies commercial to get to her judging assignments at AKC dog shows, but she doesn't share Hugh's love of flying small planes. She has enjoyed his aviation friends and never complains about the time Hugh spends in the air or in the hangar.

Now with an airplane that was well equipped but easy to fly and comfortable, he had no trouble filling the available time going to weekend fly ins, flying two hours one way for a $5 breakfast of sausage and pancakes, and even flying down to Clemson numerous times for various meetings, reunions, etc.

Some of the major periods of pleasure while flying 40N:

August, 1993- Practical flight test completed for IFR Rating.

October, 1994- Trip to Colorado Springs, (He had been stationed at Camp Carson when he came back to the states in '52.) with many stops going and coming back visiting family and new places. 32 hours in three weeks.

April, 1995 Cross country to Newberry, SC fly in. Experienced generator failure in pattern, repaired on field. Then he flew to Elizabeth, TN to visit friends. Next stop was Decatur, AL, with another generator failure enroute. IA installed a spare generator shipped from home, and as he headed home three days later the generator failed five minutes away from entering IFR conditions. He went back to Decatur and left the plane for a month while IA installed a new alternator.

September, 1997- 120/140 Convention in WVA, then on to St Louis for a long weekend as a tourist, then back home through Decatur and Clemson 20+hrs over two weeks.

November 18, 1997 He set a National Aeronautical Association speed record for aircraft class between Richmond and Clemson, 3hrs 40min, 94.8 mph w/15mph headwind

September, 1998- Cross country to convention in Chino, CA, flew out to Catalina Island, four weeks, 6000 miles, visited Alamo in San Antonio,63 hrs in left seat, July, 1999- Six weeks following Lewis & Clark Trail to the west coast. 5000+miles, 54 hrs in the air- many stops on his way back with family and friends. Visited all the Indian museums. At one of the museums he met a young Indian woman, Dr. Monica Mayer- "Good Medicine" is her Hidatsa name, who had returned to her tribe to concentrate on preserving the Indian knowledge of medicinal herbs that had never been written down.

Visited Wolf Education & Research Center in ID and their special wolf pack in the wild.

The following paragraph is from the article he wrote for one of his flying trips and is perhaps his favorite description of all the experiences and the meeting new people all over the country.

"After 11 days of flying, the 'spirit of the trip' came to the surface when I landed at Lewiston, ID. As I was putting 40N to bed, a 172 pulled into the next space with a young couple from Sacramento in for vacation. We talked at length and when the woman learned I had flown the 120 from Virginia along the L & C Trail, she jumped around and loudly exclaimed, 'I love this! I really enjoy seeing the old folks out having fun and doing things with gusto!' I took that as a compliment and a challenge to keep doing it as long as I can."

May, 2000 Began trip to Alaska. and got to Ft St Johns, BC but the weather forced him back. Visited Little Big Horn battlefield, landed at Chicago's Meigs Field, saw a Cubs game at Wrigley Field, etc. 60 hrs flying Gone 41 days, lost 15 due to weather.

April, 2001 Munich, Germany Flew as passenger in German C-140, right up to edge of the Alps, introduced to the Dolderer family who ran the FBO at Memmingen, logged flight in their French DR400, had dinner with the family.

September, 2001 After the World Trade Center destruction, visited all over New England, 18 hrs total (No flying for 15 months because of bypass, real hassle to get medical back.)

September, 2005 To Leadville, Colorado and then to Omaha and then back home, 37 hrs flying

Each one of these flights was filled with the pleasure of new places and interesting people. Flying all over the country, in all kinds of weather, landing at every size airport in an airplane with 100hp engine and 1450 lbs max weight provided a challenge that was sometimes a little nerve-wracking, but always enjoyable and educational.

Hugh has been a member of AOPA and EAA since 1968. He has served as Chap. Pres. and filled many other duties with EAA Chap 231. He thoroughly enjoyed flying in the Young Eagles program and has given airplane rides to 110 kids, introducing them to flying, one at a time. His Cessna 120, 40N was sold on March 4, 2006, and at that time Hugh had a total of 1845 hours, 1300 in 40N. Over the years he has made many flights to Clemson to participate in the Senior Platoon alumni exhibition drill team.

Hugh has been active in Clemson's proud military heritage and he helped raise more that $590,000 as fund raising chair for the Military Heritage Plaza there at Clemson and $31, 000 for the Clemson Corps ROTC scholarship program. He served as assistant fund-raising chair for the Class Of 1951 Golden Anniversary project, which collected almost $5 million for academic support. He is active in IPTAY and the Richmond, VA. Clemson Club as well as Chesterfield County/Colonial Heights Crime Solvers program. He was a founding director of Crime Solvers in 1984 and was the President in 1988. He has also been active in environmental and zoning concerns.

He is married to the former Sue Spence and they have a son, Bryan Woodle and a daughter, Ann W. Barrar and a grandson Philip Barrar.

ROBERT R. RUSSELL

Distinguished Real Estate Executive,
Professional Acrobatic Pilot, and Captain USAFR-Ret

Bob Russell became a freshman at Clemson in the fall of 1946 and while at Clemson during his sophomore year he began taking flying lessons at the nearby airport that had recently began operation in late 1946 where he later soloed a Piper J-3 Cub and began working toward his private pilots license. It was at this time he began thinking about creating a flying club there at Clemson so he was the organizer of the 3rd Clemson Aero Club, when he initiated its first meeting in his room at the College in October 1947. The meeting was attended by 30 interested cadets and returning WWII veterans. It was agreed at the meeting that a Piper J-3 Cub should be purchased if one could be found so the 21 future pilots each chipped in around $38 to jointly purchase the Piper Cub for $750. Bob found a Cub for sale in Greenville and the Club purchased the Cub. Bob then flew it back to the Clemson-Central airport. The club began its operation from this airport that had just started its operation after the operator had moved his flight training operation from the Seneca, SC. The Aero Club was officially recognized by Clemson's administration in April 1948 and Mickey Dorsey was elected President and Bob was elected Vice President. The Club was officially recognized and chartered during 1949 as a club by Clemson College. While at Clemson, Bob was also involved in other activities, which included the varsity track team, the Pershing Rifles, and Senior Drill Platoon.

Russell graduated from Clemson in January 1952 with a degree in Industrial Education and received his 2nd Lt. commission thru the AFROTC program. His years as a student may have finished in 1952 when he graduated, but his service to his country, Clemson University, and the state of South Carolina continued.

Russell entered military service in May 1952 at Donaldson AFB, Greenville SC as an aircraft maintenance officer and later that year he was assigned to Greenville, Mississippi in Class 53D for his initial flight training where he flew the AT-6G aircraft. He later was assigned to San Marcos, Texas again flying the AT-6G aircraft and then later to Waco Army Air Field , Texas for his jet pilot training where he received his jet pilot wings in July 1953. While in the Air Force for further training he attended all weather fighter school at Moody Air Force Base near Valdosta, GA, where he flew the F-94C all weather interceptor aircraft. After his training at Moody he was assigned to the Air Defense Command at New Castle, Delaware and then he served in Labrador during 1954-1955, flying the F-89S,an all weather interceptor, and the F-80's. After his release from active duty he joined the firm of Dunn and Bradstreet as a credit reporter but later he decided he wanted to start his own business. He convinced a local realtor there in Columbia to take him on as a salesman and that started his career in real estate. Later he joined the South Carolina Air National Guard (SANG), in 1956 flying the F-80's and the F-86D and later the SC Air National Guard at Congree was the first Air Guard Unit in the US to get the new F-104 which they received in 1960.The SANG was recalled to active and were sent overseas to Spain in 1961 during the Berlin Crisis. Returning from the Spain deployment the SANG then began learning to fly the F-104's which was a replacement for the F-104's. After thirteen years as a fighter pilot, Russell had met his required obligation to the Guard and on his request he was released from active duty as a Captain. He returned to his real estate business and for a few years of not flying he then decided he wanted to continue his flying as a professional air show pilot which he did for another 25 years. Performing in many air shows around the country while he still was active in the real estate business Russell and Jeffcoat Realtors, Inc which he formed in 1965 with his partner Abe Jeffcoat.

As a member of the Clemson Alumni Association, he continues to give back to the academic institution that started him down this journey. He was recently involved in the reorganization of the Clemson Senior Platoon, where he drilled in 4 half time performances over a five year period.

Nationally and locally recognized, Russell has dedicated himself to his community. As Chairman of the Columbia Town Theatre, Russell not only served on the board, but was actively involved as a participant. Russell has performed in 15 plays. Roles such as Merlin in Camelot, Pontius Pilate, and the judge in Caine Mutiny make him more than just an on-looker.

Performing as a second bass for 20 years, Russell is also highly involved in the Palmetto

Mastersingers, a volunteer male choral group dedicated to being South Carolina musical ambassadors.

In 1973, Russell was elected President of the Greater Columbia Association of Realtors, has been chairman of the vital Committee of 100, and was elected President of the Columbia Sales and Marketing Executives Club in 1975. He also serves on the Executive Committee of the Celebrate Freedom Air show in Camden and was the President of the Board of the Columbia Salvation Army. Russell is listed in "Who"s Who in America", "Who's Who in American Aviation", and "Personalities in the South." He has also been awarded the Order of the Palmetto, the state's highest civilian award. Russell is the acclaimed author of two books, one entitled *Be Your Best Self*, and the second titled *Touching the Face of God.*

Most recently, Bob Russell has been elected into the South Carolina Aviation Association Aviation Hall of Fame, a highly regarded distinction. Additionally, he was named Aviator of the Year. Other SCAA Hall of Fame inductees for 2003 included U.S. Senator Ernest F. "Fritz" Hollings and Carroll Joye. Russell is currently a member of the Experimental Aircraft Association, International Aerobatic Club, and International Council of Air Shows.

Due to his passion for aviation, Bob Russell owns his own Aviation Company, BR Legend, LLC, and owns his own planes. He recently finished building his second plane, a turbine Legend, which took three years to build. The Legend is among one of 15 planes in the nation of that caliber, as it is on the cutting edge for design, materials, and power. With a Walter turbine engine that turns 40,000 rpm and a prop turning 2,000 rpm, the Legend has tremendous torque and power and can climb at over 6,000 feet per minute and reach speeds of 350 knots. From purchasing a Piper Cub as a young Clemson student to owning a unique Legend as a successful businessman, Russell has been persistent in his passion for flying. He has logged over 8000 flight hours. Some of the aircraft the Bob flew as an aerobatic performer included a modified DeHavilland "Chipmonk," DHC-1B-253, a Marchetti SF 260 Waco "Meteor," and an extra 3005.

In 1965, Bob Russell formed his own Real Estate Company. After nearly 40 years of service in real estate, Bob Russell and his business partner, Abb Jeffcoat, have a successful business, operating 13 offices in South Carolina with over 450 sales associates. Russell & Jeffcoat Realtors, Inc. is the #1 listing and selling real estate agency in the Midlands, is recognized as one of the leading residential real estate firms in the Southeast region of the United States, and is among the top 100 real estate companies in the nation.

Russell is devoted to his wife Patty and their five children. Russell's passion for life and his community stems from years of service to his country and surroundings. Lessons that were learned almost 55 years ago in a little town called Clemson.

This bio has been modified from the bio created by Ashley Shelly who kindly provided it to the author for my use.

LT EDWARD A. SMITH

Edward A. Smith graduated in June 1951 from Clemson A & M College, Clemson, S.C. with a degree in Industrial Education. He was active in a number of college activities, was an AFROTC Distinguish Military Student, and recieved his commission as a 2nd Lt USAFR. He volunteered for immediate active duty in the USAF in June, 1951 and was assigned to MATS, Brookly AFB, Mobile, Al. He had applied before graduation for pilot training and was accepted for training as a Student Officer in Class 52 G, reporting to Spence AFB, Moultrie, Ga. in October 1951. He completed his training at Spence and was transferred to Williams AFB, Big Springs, Texas where he completed his pilot training and received his wings on 25th October 1952. He went thru additional combat crew training and was assigned to the 5th Air Force, 4th Fighter Interceptor Wing, 336th Sq, Able Flight, at Kimpo AFB, Korea in March 1953. He flew some 20 plus combat missions as an F-86 pilot and wingman and was awarded an Air Medal during his tour. He returned to the States in March 1954 and was then assigned as a 1st Lt. to the Tatical Air Command, 9th Air Force, 366th Fighter Bomber Group, 612 Fighter Squadron at England AFB, Alexanderia, La. He was married to the former Betty Turner of Spartanburg, S.C. in April 1954 and was later killed on 6 August 1954 while flying a F-86F during NightHawk Operations.

LT JOHN C. WILSON

John C. Wilson was born in Lyman, SC., the son of Audrey Howard Wilson and Marion Crow Wilson. The family moved to Greenville, SC where he attended school, graduating from Greenville High School in 1946. He entered Clemson College where he majored in Electrical Engineering. He was a member of the AFROTC and was the Commander of the Arnold Air Society. He graduated with a degree in Electrical Engineering, Class 19511951, recieving his commission as a 2nd Lt USAFR and entered military service on 10 June 1951. He was assigned to the 2156 Air Rescue TTU, Macdill AFB, Tampa, FL. He later was assigned to Pilot Training Class 52G as a Student Officer at Bainbridge, Ga. in 1951. He completed his training, receiving his silver wings, and then was assigned to Pinecastle AFB, for training at the All Weather Interceptor School. After graduating he was assigned for duty in Japan, serving there from 1953 thru 1956, where he flew the F-86'S and F-94C's He returned to the States, serving at Moody AFB, GA, at Tyndal AFB, Fl., and later at Maxwell AFB, Montgomery, Al. He then was assigned to Clovis AFB, NM where he flew the C-119's. He was released from active duty and for a time he was employed as a pilot for Hayes Aircraft in Birmingham, Al and was associated with the NASA Space Program in Huntsville, Al. He then began working for North Carolina Power and Light as Senior Electrical Engineer. He was married to Betty Joe Hester and he retired in in 1994. Tragically his wife died in September 1995 and he died in December 1995.

WILLIAM R. AUSTIN, II
Colonel USAFR-Ret

William attended Clemson University and graduated in 1959 with a BSME and he later enlisted in the Air Force in June 1960. He received his commission as a 2nd Lt after completing Officer Training School. He then attended Navigation Training at Harlingen AFB, Texas, and received his navigator wings in July, 1961. His next assignment was at Keesler AFB, Mississippi. While there, he completed Electronic Warfare Officer's School. His first operational assignment in July 1962 was with the 55th Strat Recon Wing, Forbes AFB, Kansas. While there, he was accepted for pilot training at Laughlin AFB, Texas in 1965. After receiving his pilot rating in June, 1966, he attended F4 Training at Davis Monthan AFB, Arizona. Next he completed the Wild Weasel and Survival Training courses before reporting in May 1967 at Ubon RTAFB,Thailand.

He was shot down on October 7, 1967, after having successfully completed 80 missions with the 555th Tactical Fighter Squadron (Triple Nickel). The lead aircraft, the F-4D Phantom was on a photo reconnaissance escort mission over North Vietnam and was piloted by Major Ivan D. Appleby. The aircraft was hit by a surface to air missile(SAM), losing its hydraulic pressure, and then the aircraft started to roll thus becoming uncontrollable. Major Appleby then ordered bailout. Captain Austin, the backseater, bailed out but Appleby was not seen to eject as no chute was seen by Austin. He observed the impact of the aircraft about 3 to 5 miles where he landed and was immediately captured and sent to a prisoner of war camp where he remained from October 7, 1967 till his release on March 14, 1973. After a period of medical convalesce he was returned to active duty as a Mission Ready Pilot flying the A-7D. He became a student and was a graduate of the Air Force Command and Staff School at Maxwell AFB, Montgomery, Al. Later in his USAF career he was promoted to Colonel in 1980 and was Deputy and Base Commander for five years.

His awards include the Silver Star, Distinguished Flying Cross with an Oak Leaf Cluster, Air Medal with nine Oak Leaf Clusters and the Armed Forces Expeditionary Medal.

His strong Christian beliefs and his confidence in himself and the other prisoners kept him going while in the prison camp. Confidence as he had stated "That's what kept me going while I was in prison. Confidence in God, country, family, fellow man, and self gave me the extra strength to keep going when things really got tough. I knew God was always there, and that He would answer my prayers if I made them reasonable. I prayed for strength to go on, and guidance in helping me use my abilities to outwit the enemy. After thinking about Christ's Crucifixion, I couldn't bring myself to ask God, 'Why me?' or ask God to free me. God was there, and that's a large part of why I'm here. Good old United States of America! I knew it was backing me up and trying to do what was right, and if I did my part I would be back to freedom someday. I was confident that my wife Myrtle would raise our children and take care of everything at home as near as possible to the way we would have done together. Also, I knew she would support me as she always had, and that our children would follow her example. I was sure my family, church, community and friends were behind me with prayers and other support. I had confidence in myself that I could survive and overcome almost anything with a team like that to help. Sure enough, the whole team was there--God, country, family, friends".

During 1985 thru 1988 he was the AROTC Professor of Aerospace Studies at Clemson University and later in November 1996 William Austin II retired from the United States Air Force as a Colonel with 28 years of active service in October 1988. He and his wife Myrtle reside in Oconee County outside Clemson, SC on a waterfront lot on Lake Hartwell.

COLONEL ALTON C. WHITLEY, JR.
USAF (Ret)

Colonel Alton C. Whitley, Jr., is the Director of Air Force Junior ROTC Department and an Aerospace Science Instructor with the 941st Unit of Air Force Junior ROTC located at D. W. Daniel High School, Central, SC. He assumed this position in July 1994 following his retirement from active duty. During his tenure the unit has twice earned the prestiguous "Sword of Excellence" Award which is presented each year by the South Carolina Air Force Association to the most outstanding of the 36 Air Force Junior ROTC units in the state. Colonel Whitley has also been recognized with the Air Force Junior ROTC "Outstanding Instructor" award. He is a member of Tiger Brotherhood and serves as the Secretary/Treasurer of the Board of Directors for the Clemson Corps, an alumni group devoted to preserving the military heritage of Clemson University as well providing scholarship assistance to Senior ROTC cadets.

Colonel Whitley was born November 27, 1945, in Concord, NC, and graduated from Lancaster High School, Lancaster, SC, in June 1964. He received a bachelor of arts degree in economics from Clemson University and was recognized as a distinguished graduate of the AFROTC in 1968. Colonel Whitley earned a master's degree in public administration from the University of Northern Colorado in 1975. Colonel Whitley is a distinguished graduate of Squadron Officer School and completed Air Command and Staff College in June 1973. He graduated from the National War College, Washington, DC, in June 1986. He completed the Air Force Junior ROTC Aerospace Science Instructor Certification Course at Air University, Maxwell Air Force Base, AL, in 1994.

After completing Undergraduate Pilot Training at Webb Air Force Base, TX, in June 1969, Colonel Whitley transitioned to the F-100 Super Saber. While assigned to the Republic of Vietnam in 1970, he flew 143 combat missions in the F-100 with the 531st Tactical Fighter Squadron at Bien Hoa Air Base and the 309th Tactical Fighter Squadron at Tuy Hoa Air Base.

Colonel Whitley returned to England Air Force Base, LA, and flew the F-100 with the 431st Tactical Fighter Squadron and the 416th Tactical Fighter Squadron. After completing Squadron Officer School, he transitioned to the A-7D Corsair II and was assigned to the 74th Tactical Fighter Squadron at England Air Force Base, LA, in December 1972. In February 1973, Colonel Whitley returned to Southeast Asia as an A-7D search and rescue pilot with the 3rd Tactical Fighter Squadron at Korat Royal Thai Air Force Base, Thailand, where he flew 90 more combat missions. In February 1974, Colonel Whitley was assigned to the 422nd Fighter Weapons Squadron, Nellis Air Force Base, NV, as an A-7D operational test and evaluation project officer. In September 1975, he was assigned to the 355th Tactical Fighter Wing, Davis-Monthan Air Force Base, AZ, as a wing weapons & tactics officer. In March 1976, he joined the A-10 Follow-On Test and Evaluation Team at Davis-Monthan and served as test project officer until he was reassigned to the 22nd Fighter Weapons Squadron, at Nellis Air Force Base, in April 1977.

Colonel Whitley completed the A-10 Fighter Weapons Instructor Course in 1980. He was the Outstanding Graduate of his class and later received the Risner Trophy recognizing him as the Outstanding Graduate for all weapons systems that same year. After graduation, he was assigned to the 66th Fighter Weapons Squadron as a fighter weapons instructor and assistant operations officer.

From January 1981 to July 1985, Colonel Whitley was assigned to the 4450th Tactical Group at Nellis Air Force Base, flying the F-117A and A-7D aircraft. He served as group staff officer, operations officer, and squadron commander while assigned to the 4450th Tactical Group. In this assignment, on October 15, 1982, Colonel Whitley became the first operational pilot of the Tactical Air Command to fly the F-117A stealth fighter.

In June 1986, Colonel Whitley completed the National War College in Washington, DC and was assigned to the Directorate of Plans, Deputy Chief of Staff/Plans and Operations, Headquarters Air Force, Washington, DC, where he served as Deputy Division Chief of the Tactical Forces Division, Division Chief of the Munitions Planning Division and chairman of the War Readiness Materials panel. In August 1987, he was assigned to Nellis Air Force Base as commander of the 4440th Tactical Fighter Training Group (Red Flag). During this time, he completed the Adversary Tactics Instructor Course and flew the F-5E attached to the 65th Aggressor Squadron.

In July 1988, Colonel Whitley became vice commander of the 57th Fighter Weapons Wing, Nellis Air Force Base, flying the A-10 with the USAF Fighter Weapons School and the 422nd Test and Evaluation Squaron. He was assigned to the Operations Deputate of Headquarters Tactical Air Command in July 1989, where he served as Director of Fighter Operations and subsequently as Director of Fighter Training and Tactics. Colonel Whitley commanded the 37th Fighter Wing from August 1990 until its deactivation in July 1992. During Operation DESERT STORM, Colonel Whitley commanded the 37th Tactical Fighter Wing (Provisional) and flew 19 combat missions in the F-117A stealth fighter.

Colonel Whitley retired from active duty on June 30, 1994, after over 26 years of service to his country. In his last active duty assignment, he served for two years as the Professor of Aerospace Studies and Commander of Air Force ROTC, Detachment 770, Clemson University.

A command pilot with more than 5,500 flying hours, Colonel Whitley has been awarded the Legion of Merit with two oak leaf clusters, the Distinguished Flying Cross with four oak leaf clusters, Meritorious Service Medal with two oak leaf clusters, Air Medal with fourteen oak leaf clusters, Aerial Achievement Medal, Air Force Commendation Medal with one oak leaf cluster, Army Commendation Medal, Armed Forces Expeditionary Medal, Southwest Asia Medal with two bronze stars, Vietnam Service Medal with three bronze stars, Republic of Vietnam Gallantry Cross with palm, Republic of Vietnam Campaign Medal, and the Kuwait Liberation Medal.

Colonel Whitley is married to the former Mary Ann Arrowood of Union, South Carolina. Ann is an academic advisor for education majors at Clemson University. They have two children: a son, A. C., who earned both his bachelors and masters in accounting from Clemson and a daughter, Elizabeth, a Clemson graduate in Spanish and International Trade who currently serves as an intelligence officer in the United States Air Force.

GEORGE VENTURELLA

USAF Pilot and Airline Captain

George entered Clemson and after 4 years he was graduated in 1982 with a BA degree in Political Science . He was very active in student affairs and in the AFROTC. He was the Arnold Air Society Commander 1980-81, and the Cadet Vice Wing Commander, Cadet Colonel 1981-82. He was also one of 10 students from Clemson named to Who's Who Among Colleges and Universities. After graduation as a 2nd Lt he attended pilot training at Columbus AFB, MS Class 83-08. After completing his training and receiving his silver wings he then was assigned to the 4450th TAC Group to fly the Mitsubishi MU-2 based at Nellis AFB, NV. The 4450th TAC Group was the operational cover for the stealth F117 which at that time was a classified aircraft. The cover operational aircraft was the A-7. He would fly from Nellis AFB to Tonopaugh Test Range, and other remote desert locations where the F-117 was to be operated. They flew the MU-2 in civilian paint scheme, civilian call sign, and civilian clothes. When to MU-2 lease expired in 1985 the USAF brought in 4 C-12F, King Air 200B, and had them painted in Civilian colors. Same mission, different aircraft. The C-12 was attached to the 1400 MAS, and they were Detachment 5, 1986-89. He flew C-130E with the 40th Tactical Airlift Squadron, Pope ABF, NC. During December 19, 1989 his squadron dropped troops of the 82nd airborne on Rio Hato AB, Panama during the invasion. George got to remain back at the base for the birth of his son at 9:42 pm. He requested and was released from active duty. In 1990 George flew for USAir based in Charlotte, NC as a 727 first officer. During 1992-96 he flew the NC Air National Guard C-130H. He has been employed by American Airlines since 1991. The AA aircraft George has flown as Captain was the MD-80. First Officer A-300, B-777, B-767, B-757, MD-11. His type ratings:L-382, MD-11, A-300, B-777, B-767, B-757, MD-80.

DR. EDWARD F. BYARS, PhD

Engineer, Educator, Antique Aircraft Restorer
and General Aviation Pilot, and Glider Pilot

Ed was born in Lincolton, NC and his family moved to Charlotte, NC first and later to Rock Hill, SC. He entered Clemson in 1942 and was very active in the student body and in the Army ROTC Cadet Corp as Private, Sergeant, First Sergeant, Captain, Member of Tiger Platoon 2,3,4, Executive Sergeant's Club 3, Officers Club 4, Clemson Dance Association Business Manager 4, York County Club 1, ASME 3,4.

In 1946 he received his BSME degree. He entered military service in 1946 as a member of the US Army Signal Corp. While at Clemson and as a faculty member during 1947 thru 1960 he earned a Master's Degree in Civil Engineering and in 1957 he received his PHD in Engineering Mechanics from the University of Illinios.

He left Clemson in 1960 to become Professor and Chairman of Mechanical Engineering at West Virginia University. He became the Dean of Engineering there in 1979. He returned to Clemson in 1980 and served as Executive Assistant to the President until he retired in 1985.

He is Professor Emeritus at Clemson University and West Virginia University.

During his professional career he authored several engineering textbooks and numerous research papers and articles, and developed several successful patents. Many universities have used his widely acclaimed text books. He has also written several books about his hobby of flying and soaring.

After his retirement he became the business manager for his wife Betsy, a noted author of children's books, and he pursued his hobby of competition soaring and restoration of antique aircraft.

He began his aviation career when he soloed a J-3 Piper Cub while still a Clemson student, taking his flying lessons from Ben Johnson at the Anderson Airport, Anderson, SC. He bought his first airplane, a 1931 Stinson SM8A, in 1949. Since then he has owned over 60 aircraft including 18 Gliders, a Money Mite, two Culver Cadets, two Fairchild 24Ws, several light twins, several open cockpit and cabin Wacos, and several Aeronca Champs and Piper Cubs. He still has a Beech Bonanza that he has owned for at least 26 years that he uses for serious travel. He also has a J-3 Cub that he has owned for 33 years as well as a couple of antiques and a sport biplanes.

He has raced high performance sailplanes for 43 years till a serious accident in 2003. He is a life member and was Chairman of the Board of the Soaring Society of America and was indicted into the Soaring Hall of Fame in 1992. He has written various articles for magazines about flying and soaring as well as several aviation books.

He holds the following FAA ratings as Commercial Single and Multiengine, Glider, Instrument, Single Engine Sea as well as an A&P license.

He resides at a private airport, Eagle Ridge(SC24) which opened in 1993 about two miles west of the present Oconee Regional Airport (CEU). There are five other families at this private airport.

So you see that Aviation has been a very important part of this individual's life.

GENE MCCOY

An Overview Biography

He is a seventh generation low country boy gentleman and farmer.

Gene graduated from Clemson College, Class 54 Civil Engineering, and received his USAF commission thru AFROTC. Entered the USAF 1954 as 2nd Lt assigned to Wright Patterson AFB and was an F-104 Flight Test Engineer, a back seater(GBI)F-104 engineering testing. Released from the USAF and went to work for NASA.

He held the following NASA Management and Engineering positions beginning in the early Manned Space Program to the Shuttle Program.

Mercury Program Period-

Pad Coordinator -Pad 5 (Ham, Shepard, and Grissom launches)
Senior Launch Complex Engineer(Glenn,
Carpenter, Schirra, and Cooper launches)
Deputy Chief, Ground Systems Division
Gemini Program - Spacecraft Test Conductor

Apollo Program - Chief, Manned Spacecraft Office
Skylab Program - Chief, Space Vehicle Office
Shuttle Development -Chief, Space Shuttle Payload Integration Office
Shuttle Operations - Chief, Future Program Office

During his NASA employment Gene gave many one to one briefings on"Launch Pad Operations for Manned Space Flight" to a number of individuals including Alan Shepard, Gus Grissom, John Glenn, Frank Borman, Neil Armstrong, the Chairman of Exxon, and the Duchess ofLuxembourg.

He was for 24 years an Adjunct faculty member with the Florida Institute ofTechnology.

Gene has conducted numerous seminars, workshops, and banquets programs on Management, Strategic Planning, and Leadership over the years. He was a Sloan Fellow at the Stanford Graduate School of Business under Dr. Porterfield many years ago and took the course again 27 years later and got the same grade.

He designed a number of mathematical models of the common stock for one hundred major Corporations, maintained them, and published a privately circulated stock market news letter.

Gene did consulting work for Boeing, United Technology , Rockwell, and Northern Telcom, among others, and still is active although retired.

He has been responsible for the operational and financial management of three small corporations, one general partnership and eight investment trusts.

He cleared US Customs as a Captain of a British sea going vessel, became an experimental aircraft builder, a Glider, Sea Plane Pilot, and was a 50% owner of a Florida Marina for 27 years.

Gene traveled extensively through the US(50 states and 21 countries)
including Labrador, and Alaska twice in his pickup camper.

EDWARD J. HUTTO, JR.

Electrical Engineer and Space Researcher

Ed Hutto Jr attended Clemson as a student in September 1940, majoring in Electrical Engineering, but military services called and he became a member of the Army Signal Corps as an officer and served in the 3168th and 4025th Signal Service Battalions in the Philippines Islands, Luzon. After the war he returned to Clemson in 1946, graduating with a degree in Electrical Engineering in February 1947. While working toward his degree he was on the SLIPSTICK staff, a member of Phi Eta Sigma, Phi Kapp Phi, and Tau Beta Phi.

After he was graduated he was employed by the Radio Corporation of America (RCA) in their student training program in Camden, NJ. After training in this program he then began working in the Optics group headed by I.G. Maloff, co -author of "Electric Optics in Television" which was the first text on this subject. Ed participated in the design of a three color projection receivers used in the Federal Communication Commission (FCC) hearings which ruled against the Columbia Broadcast System(CBS) color system that use a rotating filter system. Since the CBS system was incompatible with the existing black and white receivers, RCA was the winner and all-electronic television transmissions systems would be developed by RCA. During 1953 he began working on a TV camera optical system using image orthicon sensors that was being developed for the USAF at Wright Patterson, AFB pre-Sputnik. They tested it by photographing images during the construction of the Walt Whitman Bridge which was in line of site from their lab window in the Camden area looking over toward South Philly. His next job later in his career with

RCA was as Project Engineer for the Mark I Solar Simulator that was later installed at the USAF's Arnold Engineering and Development Center, Tullahoma, Tenn. Later on he was working with the RCA Aero-Electronics Division in Hightstown, NJ performing analysis of the ground display system for the lunar impact probe, the Ranger Spacecraft, being developed by the Jet Propulsion Labs, Pasadena, CA. Later in his career he selected and tested the fish-eye lens for the TIROS I camera for use in the first NASA Weather Satellite System launched April 1,1960. This was a TV and Infrared Observation Satellite and it returned 22, 952 photos till it stopped transmitting June 17,1960. He was Project Engineer for the Laser voice transmitter system for the Gemini VII and later he was part of the RCA team that evaluated expected imagery performance of the Charge Coupled Devices (CCD) that were being developed for use in NASA Earth and Military Classified Satellite Observation Systems.

He is a member of the Optical Society of America (OSA) and has made presentations and has published articles for the OSA. He has also developed articles for publication in the Society of Motion Picture and Television Engineers (SMPTE) Journal and co-authored articles for the RCA Corporate Engineering Service publications.

He is a member of the Association of Recorded Sound Collections and the International Association of Jazz Record Collectors. He has produced a number of multi-slide shows about artists of the swing era. The Library of Congress now has his production of "Eldridge Johnson and the Victor Talking Machine Company" materials. The Talking Machine Company was purchased by RCA in 1929 to give it another manufacturing facility. Ed Hutto is probably the best authority about this company and its operations as well as information about Eldridge Johnson, it founder. He also produced a slide show presentation, "The South Carolina Railroad" for the 100 year anniversary of Dorchester County, SC in 1997 and also for presentation at the Branchville Railroad Museum. He retired from RCA in 1984, residing with his wife Mary and three daughters in St George and Orangeburg County, SC. He is now resides at Accomac, Virginia, on the eastern shore of Va.

WALTER A. SMITHWICK

Major USAF-Ret Reconnaissance Pilot

Walter as a young person built model airplanes and this set the stage for him to aspire to be a pilot. He was Civil Air Patrol Cadet in 1944 when he got his first flight in an aircraft. After finishing high school he enrolled in Clemson in 1945 to study for a career in Diary, receiving his BS Degree in 1949. While at Clemson he got a number of flights with WWII pilots who flew out of the Clemson-Central Airport. He entered the USAF as an Aviation Cadet in 1950 was assigned in 1951 as an aviation cadet to USAF Pilot Training 52H at Kinston, NC. He completed this phase of training and was assigned to Perrin AFB, Texas and then to Craig AFB, Al to complete undergraduate flight training, receiving his silver wings in December 1952. He then went to all weather Interceptor School at Moody AFB, Valdosta, Ga. flying the T-33 and later to Tyndall AFB, Fl flying the P-80A for combat crew training in the F-86D. He was then sent to George AFB, California where he was assigned to a Fighter-Bomber Wing, flying the F-86F.

The next training was as a ground controller at Tyndall AFB. From there he was assigned to England AFB, La., Donaldson AFB, SC, Fort Bragg, NC. and then to Langley AFB, Va. where he and his team manned ground radar equipment supporting Tactical Air Force operations. He worked with big mobile computers and auto tracking radar guiding and controlling aircraft for precision munition drop.

During this phase of his career he assisted in Atom Bomb testing in the Nevada Desert, himself being selected twice as a "guinea pig" during bomb test. He spent a year north of the Arctic Circle at a Dew Line radar site, then as an instructor at Tyndall in ground control operations. He returned to flight operations as a flight instructor at Williams AFB, Arizona instructing in the T-33 and T-37. He volunteered to fly the RF101Voodoo during the Vietnam War flying quite a number of exciting recon missions. As he stated to me "The Reconn credo: Alone, Unarmed, and Unafraid was most appropriate. Well he added two out of three isn't too bad." It was quite an exciting experience flying the "101 at six hundred miles per hour, visual only navigation, hearing the acrimonious twitter from the hostile radar through the headset, with a few of the protective services provided by the Strike Forces. All this plus an aircraft afterburner takeoff, in-flight refueling, and the drag chute braking after touchdown during landing. After completing 49 missions as a recon pilot he later became a Staff Officer at 7th AF Headquarters in Siagon where he prepared target assignments as directed. He returned to the states and returned to flight training duties in the T-38 at Craig AFB, Al. He retired as a Command Pilot with over 5000 hours of flight time in a large radial engine aircraft, jet powered aircraft, laison aircraft, and helicopters.

His Military Decorations:

Air Medal with 4 Oak Leaf Clusters
SO-G-324, Headquarters 7AF, 67
Presidential Unit Citation
National Defense Service Medal w/1BSS
AF Longevity Service Award w/3 Oak Leaf Clusters
Good Conduct Medal
Armed Service Reserve Medal
Vietnam Service Medal w/1BSS
Republic of Vietnam Campaign Medal
AF Outstanding Unit Award
Small Arms Expert Marksman Ribbon

LT BEN INGRAM

USAFR Strategic Missile Crew Operations

Ben was born December 21, 1938 at Pageland, SC, graduated from high school and entered Clemson in 1957 to become a mechanical engineer and received his BS ME degree in 1961. While attending Clemson he was active in the Glee Club, ASME, and SAE Clubs. He entered military service, was graduated fron AF Officers Training School (OTS) and was commissioned a 2nd Lt and in 1962 reported for duty to Davis-Montham AFB, Tucson, AZ. He was then assigned to Sheppard AFB, Wichita Falls, Texas where he was graduated from Missile Launch and Operations School, and attended Launch Operations School at Vandenburg AFB, CA. While at Davis-Monthan AFB he worked as a member of the Site Activation Task Force (SATAF) supporting Titan II site construction and later Missile Crew Operations. He remained on duty there till 1965 and was released from active duty and placed in reserve. He began his career in industry and worked for the following industries.

Owens Corning Fiberglas 1965-1970 Engineer-Scientist , Granville,OH
Reichhold Chemicals Inc 1971-1989 VP Marketing & Research, Nashville, TN
He later formed his own Real Estate Appraisal company, the Ingram Company in Greenville, SC.

He and his wife have two sons, Mark who works for the Merck Corp, Albany, GA and Brad who works for CNN, Atlanta). Both sons are Clemson graduates (Mark-BS Chemical Engineering & Brad- BS Accounting). Ben and his wife also have two grandchildren (boys) in Atlanta, Ga.

COL HOUSTON TERRY HAWKINS

USAF (ret)

Director Nonproliferation and International Security

Los Alamos National Laboratory, Los Alamos, New Mexico

MAJOR RESPONSIBILITIES:

Management of a staff of 900 scientists, engineers, support staff, and contractors tasked with developing applied scientific and engineering methods to detect, deter, and counter the proliferation of weapons of mass destruction and the use of such weapons by terrorists. Oversee scientific programs related to counter-terrorism and counter-narcotics; to assessment of foreign technologies; to export control and the safeguarding of nuclear materials; to advanced detection technologies; to information assurance and cyberwarfare; and to the exploration of outer space. Program Manager for Special Access Programs. Director LANL Field Intelligence Element.

PRIOR EMPLOYMENT/DUTIES:

1988-1993 Division Leader, International Technology Division, Los Alamos National Laboratory, Los Alamos, New Mexico (1988-1993); Deputy Director for Arms Control & Test Limitations, Defense Nuclear Agency, Washington DC (1987-1988); Special Assistant for Air Force Nuclear Matters, Office of the Secretary of Defense, OATSD(AE)/ Pentagon, Washington DC (1983-1988); Chief, Nuclear Weapons Branch (and Acting Chief, Nuclear Energy Division --1982-1983), Defense Intelligence Agency, Washington DC (1979-1983); Assistant Division Chief, Radiometrics Division, Air Force Technical Application Center, Patrick AFB, FL (1974-1979); Chief, Production Control Unit, McClellan Central Laboratory, 1155th Technical Operations Squadron, McClellan AFB, CA (1968-1974); Chief, Reconnaissance Branch, 18th Armament and Electronic Squadron, 18th Tactical Fighter Wing, Kadena AB, Okinawa (1967-1968); and Staff Chemist/Materials Scientist, Air Force Materials Laboratory, Wright-Patterson AFB, OH (1963-1967).

EDUCATION:

Clemson University, Clemson SC, Distinguished Military Graduate, BS Chem, 1959-1963.
R&D Laboratory Management, Air Force Institute of Technology/Ohio State, Dayton OH, 1964.
Nuclear Research Officer Certification, (Research Topic Auto-Redox Reactions in Aqueous Solutions of Plutonium), McClellan Central Laboratory, McClellan AFB, CA.
La Salle Extension University, Chicago IL, American Business Law, 1969-1974.
Air Command and Staff College, Maxwell AFB, AL, 1976.
Air War College/Aerospace Management, Maxwell AFB, AL, 1977.
National Defense University/National Security Management, Washington DC, 1978.

SPECIAL TRAINING:

Applied Microscopy, McCrone Institute, Chicago IL.
Thermal Properties of Plastics, UCLA, Los Angeles CA.
Plastics Technology, Stephens Institute of Technology, Hoboken NJ.
Catalysis and Kinetics, MIT, Cambridge MA.
Precision Reconnaissance Photography, ATC, Lowery AFB, CO.
Senior Officer Nuclear Weapons Course, Kirtland AFB, NM.

PANELS/COMMITTEES:

Los Alamos Information Security Policy Board, 2000
Science & Technology Intelligence Committee, 1997
Joint Atomic Energy Intelligence Committee, DIA Representative, 1979-1983
Nuclear Underground Test Monitoring Working Group,
DIA Representative, 1979-1983
President's Verification Review Panel, 1986-1988
Interagency Hard Target Kill Working Group, Co-chair with
MGen Norm Wood, 1986-1988
Atmospheric Obscuration Panel, Executive Secretary,
Defense Science Board, 1986-1987
Nuclear Defense Agency Reorganization Panel, Executive Secretary,
Defense Science Board, 1987
National Defense Information Control Advisory Panel,
OSD/OUSD(R&E) Representative, 1985-1987
Scientific Advisory Board, Center for Microelectronics,
University of South Florida, 1993-1996

HONORS:

Laurel Award, Aviation Week & Space Technology, 2000
Distinguished Business Management Award, Los Alamos
National Laboratory, NM, 1999
Legion of Merit, 1988.
Defense Superior Service Medal, 1987.
Defense Superior Service Medal, 1983.
Air Force Meritorious Medal, 1979.
Valley Forge Freedom Foundation Medal, Public Speech, 1993
Valley Forge Freedom Foundation Award, Public Speech, 1978.
Valley Forge Freedom Foundation Award, Written Paper, 1976.
Air Force Meritorious Medals, 1974 and 1973.
Humanitarian

FRANK KELLERS III

WW II Veteran and Space Pioneer

Born in Clinton, SC, Kellers grew up traveling the world as an "army brat." The son of a Clemson graduate and colonel in the Army Signal Corps, he attended first grade in Missouri, eighth grade in the Philippines and 12th grade in Germany, but to this day he still thinks of South Carolina as home.

After graduating from Clemson with a B.S. in electrical engineering in 1957, he served in the Army Signal Corps where he met and married his wife Shelia Stalk. He returned to civilian life in 1959 and went to work for the Lockheed Missiles and Space Co. in Sunnyvale, Calif. The Kellers raised three daughters, Diana, Melanie and Lisa, still make a home there.

During his 32-year career with Lockheed, Kellers distinguished himself as an expert in telemetry, tracking and command systems in the chief systems engineer's office. He was lead engineer in the MIDAS Payload Display area. The MIDIAS satellite was developed by the USAF to detect missile launches when the missile had reached 50,000 feet into their trajectory.

When Kellers retired in 1991, Clemson became his full-time profession. President of the Northern California Clemson Club for the past 21 years, he's also served as

Alumni National Council district director, district member and member-at-large; IPTAY representative and Western regional chairman; and member of the Clemson Admissions team.

In addition, he sponsors the Clemson women's indoor track Most Valuable Player trophy and manages the Frank Kellers Annual Computer Science Scholarship for TigerNet. The Kellers routinely open their home to Clemson athletic teams. He even established the California Hotline, (408) CLEMSON, to keep everyone informed of Clemson activities happening on the West Coast.

JOE BLANDFORD

Angel Flight Pilot and Miltec Corporation Executive

Joe Blandford graduated from Clemson in June 1958 with a BS degree in Ceramic Engineering. He also received his commission as a 2nd Lt USAFR thru AFROTC. While at Clemson he was active in various the college activities and he joined the Aero Club in April 1957. He soloed the club's Funk aircraft on September 23,1957 at the Anderson Airport, Anderson, SC. After graduating he entered active duty at Lackland AFB, and then assigned to Pilot Training Class 60C at Graham AB, Marianna, Fl., where he learned to fly the Beech T-34's and Cessna T-37's. He graduated and was then assigned to Webb AFB for furthur training. He was subsequently assigned to Wright Patterson AFB,Ohio. He remained on active duty for about 3 years there at WPAFB.

Released from active duty he then became a civilian employee of Wright Patterson AFB and joined the Air Force Reserve where he served for 5 years. While at Wright Patterson he joined the Wright Patterson Aero Club, obtained his Commercial, and his Certified Flight Instructor Certificate (CFI), and instructed for the club. He later transferred to another civilian job with Engelhard Industries and during this time he was a part owner of a Cherokee Arrow. Over the years he has kept his CFI active and added single engine (SEL),multiengine (MEL) and Instrument ratings to his pilot qualifications. Joe has owned two Bellanca Super Vikings, and he now owns a Piper Seneca, a multiengine aircraft. This aircraft has been used for a number of transport missions for the Angel Flight Southeast program. The Angel Flight is a non-profit organization that flies people with sick children or family members to medical facilities or to a better place for care giving. Joe is the Big Bend Wing Leader for Angel Flight Southeast. He became interested in flying gliders so he began training to fly gliders and then added Gliders to his Commercial and Flight Instructor Certificates. He is an active member of the Rose City Soaring Association of Thomasville, Ga. He and his wife Marilyn reside in Tallahassee, Fl.

LT COL ANTHONY W. ALDEBOL

Anthony was graduated from Clemson in 1968 with a BS Mathematics. While at Clemson as a member of the AFROTC he received his Private Pilots license thru the Flight Instruction Program(FIP) and was commissioned a 2nd Lt upon graduation. He entered the USAF and became an Air Force Navigator. After the award of his Navigator's silver wings he served in various locations in the USAF as an aircrew member and later became a instructor in B-52 navigation systems. He also completed a number of extensive training courses related to weapon systems and performed over 3000 hours of flight tasking in support of Air Force objectives during the period of 1968 thru 1983.

From 1983 thru 1986 he was stationed at Barksdale AFB, La where he was Department Supervisor for aircrew evaluation. Here he supervised a team of over 30 air crew personnel who conducted evaluations of Air Force flying procedures.

From 1986 thru 1988 he was the Chief Executive Officer for the Tri-State recuiting operations. Here he supervised over 80 personnel in attaining recruiting goals covering an area of 96, 000 square miles. The annual budget was $750,000, utilizing 30 officers and a fleet of 96 vehicles.

He was graduated from Texas Christian University in Fort Worth, Texas with a Master of Liberal Arts in 1986 while still in the USAF.

From 1988 thru 1991 he was stationed at Chanute AFB, IL where he was the Chief Operations Office for the Midwest Recruiting Operations, directing admistration, logistics, advertising, and monetary resources in support of USAF Recruiting Operations in a seven state area. He managed a budget of $4,500,000 in support of over 600 recruiting personnel in 200 offices utilizing 300 vehicles.

From 1991 thru 1993 while stationed at Chanute, he was the Commanding Officer of the Technical School. Here he supervised 300 highly qualified personnel in both technical and military training of over 25,000 students anually. He was responsible for course content and teaching methodology for five major speciality technical areas. He directed resources to properly house, feed, and train student personnel.

In 1994 he completed 30 years of military service and retired to civilian life. He later became the Customer Service Manager and Military Awards researcher for the US Medals Company located in Parris Mt, Greenville, SC and is still active in that company.

TOM REESOR

Jack of all trades

Tom joined the Clemson Aero Club late near the end of the 2nd semester of 1956. The next semester the club had bought the Aeronca L-3 N49546. Tom started lessons at the Clemson airstrip between Clemson and Central. Tom's 1st instructor was Carl Bostic, his next lesson was under R. M. Taylor, and the rest of his instructions were under Grady Dalton, who soloed Tom in Feb 57. The Aero Club had a hard time getting instructors; Tom could take lessons only once or twice a month. The plane was flown to the Anderson Airport for maintenance, and it was there Tom made his 1st solo 9 Feb 56.

Tom entered the USAF as an aviation cadet but before he could start his pilot training the air force didn't need pilots so the pilot training program was terminated and the cadets were assigned to Navigator school at Hardington, Texas. However the Air Force's budget was reduced about this time. Navigators were not needed so the cadets reverted back to an enlisted grade of Airman first class. He remained on active duty for 3 years and during this time he joined the local Air Force Flying Club and obtained his Instructors certificate and instructed for their club. He also got his commercial license during this time . By 1963 he had completed his required commitment with the USAF, was released and began flight instructing and other flying jobs as they became available.

Tom was a flight instructor in Tampa starting in 1963, after leaving the AF. In 65 he went to work for Ferry Service Co., an aircraft delivery co., with offices in Wichita, Kansas (Cessna, Beech, Lear, & Helio Courier), Lock Haven, PA (Pipers except Cherokees, & Grumman AgCat),& Merritt Island, FL, (Cherokees & Aero Commander singles). His longest flight delivery was a Thrush Commander crop duster from Albany, GA, to Tel Aviv, in winter, & dusters have no heater. The engine, a rebuilt 600 hp R-1340 (probably from an AT-6), blew up taking off from London; Tom barely made it back. The rebuilding shop in Houston sent him another engine. After delivery the Thrush was involved in the 1st midair collision in Israel; nobody hurt as they were on close final, at Herzleah Airstrip.

Tom worked a Merritt Island during the Apollo days, seeing all the moon launches except Apollo 15. He taught multiengine training in aerospace writer Martin Caidin's Apache. Tom has flown in all 50 states as well as many foreign countries. Tom has some 10, 000 hours logged in 4 categories of aircraft.

Other active members of Clemson Aero Club that Tom remembers were Pete Norris, a married veteran who was club president; Russell Davis, Reid Garrison, who Tom ran into several times, the 1st was in 1964 at Atlanta Fulton Co., Apt, ferrying an AT-6 from Tampa to Danville, IL, and the last was at Anderson Airport about 10 yrs ago; and Donald Burns Clark, Private license no. 1404940, who was the club president for a year. In the mid 60s a Donald Clark was killed with his family in a C-172 in the Florida Keys. Tom says he doesn't think this was the same person.

The Aero Club had several non-injury accidents. Clark and the school physician, a pilot, got too low on final, clipping a tree and damaged a plane while trying to land at the new Clemson Cherry Bottom Airstrip down in the river bottoms (later flooded out by the Hartwell Dam). Grady Dalton, a mechanic/pilot, made temporary repairs and flew it to Anderson for permanent repair work.

On 9 February 2007, Tom's 50th anniversary, he soloed a C-172 at Conway, SC. Tom has aeronautical ratings as: Airline Transport Pilot 1479539:Airplane single and multiengine land, Rotocraft helicopter.

Commercial Pilot: Single engine and multiengine sea, Rotorcraft gyroplane, Glider, LTA free balloon.

Gold Seal Flight Instructor: Airplane Single engine and multiengine, rotorcraft helicopter and gyroplane, glider, instrument airplane and helicopter

Ground Instructor: Advanced and Instruments.

Tom has had have 2 major jobs: Aviation & TV broadcast Engineering. The current system of stereophonic TV, adopted by the FCC in 1984, was based on one of his patents which expired several years earlier.

LT COL FRANK KENT "BUCK" ROGERS

Lieutenant Colonel Frank Kent "Buck" Rogers is the Chief of Special Projects, 28th Test and Evaluation Squadron (TES), 53rd Test Management Group, Eglin Air Force Base, Florida. As chief, Lt Col Rogers directs tests of special access program advanced technologies. He is an Air Combat Command decision-making liaison and operational expert on over 20 capabilities. He also coordinates with SAF/AQ, AFMC system program offices, JCS and intelligence agencies to define requirements and develop concepts of operation for combatant commanders in conventional and information warfare. Lt Col Rogers plans and directs FDEs, technological development, and evaluations of emerging weapon systems. Prior to assuming present duties at Eglin AFB, he was Operations Officer for Detachment 1, 53rd Test and Evaluation Group, Holloman AFB, NM. Detachment 1 maximizes U.S. Air Force and F-117 combat capability by conducting Operational Test and Evaluation, Tactics Development and Evaluation, FME and reliability and maintainability testing. It also provides Headquarters, U.S. Air Force, major commands, and industry with F-117 low observable stealth tactical and technical expertise for use in weapons development.

Lieutenant Colonel Rogers entered the Air Force in May 1982 after graduating from Clemson University in South Carolina with a Bachelor of Science Degree in Physics. His military career began when he was commissioned a Second Lieutenant through the AFROTC. He has served as a combat mission commander, combat package commander, combat flight lead, evaluator check pilot, instructor pilot, and operational test pilot. Lieutenant Colonel Rogers is a command pilot with more than 4,500 flying hours, primarily in the F-117A, F-111A/D/E/F, EF-111A, and T-38A. During his air force career, he made several deployments overseas in support of Operation DESERT SHIELD/ DESERT STORM, Operation DESERT FOX, and Operation ALLIED FORCE. As an F-111 combat pilot, he earned one Distinguished Flying Cross and three Air Medals in support of Operation DESERT STORM in Iraq As an F-117 combat pilot, he received his second Distinguished Flying Cross and fourth Air Medal for combat missions he led in support of Operation ALLIED FORCE.

Lieutenant Colonel Rogers is married to the former Rhonda Levacy of Riverside, CA. They have a daughter, Megan and two sons, Camden and Connor.

EDUCATION:

1982 Bachelor of Science Degree in Physics, Clemson University, Clemson, SC
Awarded Air Force Pilot Wings Euro-NATO Joint Jet Pilot Training, Sheppard AFB, TX
1989 USAF Squadron Officer School, Maxwell AFB, AL
1993 Air Force Flight Safety Officer Course, University of Southern California
1994 National Test Pilot School, Mojave, CA
1996 Master of Science Degree in Aeronautical Science, Embry-Riddle Aeronautical University, Eglin AFB, FL
1996 USAF Air Command and Staff College, Correspondence Course, Eglin AFB, FL

ASSIGNMENTS:

Aug 82-Dec 83, Student Pilot, Euro-NATO Joint Jet Pilot Training (T-37and T-38), Sheppard AFB, TX
Dec 83-Mar 84, Fighter Lead-In Training (AT-38), Holloman AFB, NM
Mar 84-Sept 86, Aircraft Commander, F-111D, Cannon AFB NM
Sept 86-Mar 90, EF-111A Instructor pilot, Stan/Eval pilot, FCF pilot, and platform instructor, Mountain Home AFB, ID
Mar 90-Jan 93, Chief of 20th FW Stan/Eval (F-111E), flight commander, and flight lead for 19 combat missions in Operation DESERT STORM, RAF Upper Heyford, UK
Jan 93-Jun 96, Developmental Test Pilot (F-111A/E/F and EF-111A), Chief of AFDTC Flight Safety, and distinguished graduate of AF Flight Safety Course, Eglin AFB, FL
Jun 96-Jun 97, Assistant Chief of Combat Plans, 607 CPS, Osan AB, ROK
Jun 97-Aug 99, Assistant Operations Officer, 9th Fighter Squadron (F-117A) and Chief of 49th FW Quality Assurance, Holloman AFB, NM
Aug 99-Jan 02, Instructor Pilot and Chief of Training, 7th Combat Training Squadron (F-117A and T-38), Holloman AFB, NM
Jan 02-Jun 04, Operations Officer, Det 1, 53rd TEG (F-117), Holloman AFB, NM
Jun 04-Sep 05, Chief, F-117 Air-to-Ground Weapons System Evaluator, 86 FWS, Eglin AFB, FL
Sep 05-Present, Chief, Special Projects Branch, 28 TES, Eglin AFB, FL
May 06-Sep 06, deployed to Central Command, MacDill AFB, FL as Time Sensitive Target Cell Chief

FLIGHT INFORMATION:

Rating: Command pilot
Flight hours: 4,500+
Combat hours: 93
Aircraft flown: F-111A/D/E/F, EF-111A, F-117A, T-38, A/T-38, F-15B/D, F-16B/D, and T-37

MAJOR AWARDS AND DECORATIONS:

Distinguished Flying Cross (1 OLC) Kosovo Campaign Medal (1 star)

Meritorious Service Medal (4 OLC) Global War on Terrorism Service Medal
Air Medal (3 OLC) Korean Defense Service Medal
Aerial Achievement Medal (1 OLC) Armed Forces Service Medal
Air Force Commendation Medal AF Overseas Ribbon Short
Joint Service Achievement Medal AF Overseas Ribbon Long
Air Force Achievement Medal AF Longevity Service (4 OLC)
Joint Meritorious Unit Award Small Arms Expert Marksmanship Ribbon (1 star)
Air Force Outstanding Unit Award (2 OLC) Air Force Training Ribbon
Combat Readiness Medal (4 OLC) NATO Medal
National Defense Service Medal (1 star)Kuwait Liberation Medal
Kingdom of Saudi Arabia
Armed Forces Expeditionary Medal Kuwait Liberation Medal Government of Kuwait
Southwest Asia Service Medal (1star)

EFFECTIVE DATES OF PROMOTION:
Second Lieutenant 7 May 82
First Lieutenant 10 Aug 84
Captain 10 Aug 86
Major 1 Sep 94
Lieutenant Colonel 1 Mar 99
(Current as of Feb 2007)

GARLAND F. BREWER
Lockheed Staff Engineer

He was born October 28, 1933, the son of Mr and Mrs Gus W. Brewer of Seneca, SC.

He attended Seneca High School, Graduated June 1951.

He married Bobbie Wilson, April 1955. They have a son, granddaughter and grandson.

He entered Clemson A & M College (now Clemson University) in 1951 and was active in the Phi Eta Sigma, the American Society of Mechanical Engineers, and the AFROTC. He graduated June 1955 with a Bachelor of Mechanical Engineering and received his commission as a 2nd Lt USAFR.

He joined Lockheed Georgia Company (now Lockheed Martin) Marietta, GA, June 1955 and worked there till October 1955, as an Associate Aircraft Design Engineer in C-130 Project Engineering.

Garland entered the U. S. Air Force in October 1955 and began his basic flight training as a Student Officer at Stallings AFB, Kinston, NC, flying the PA-18 and AT-6g Texan. His Class 57 E was the last class to fly the AT-6G aircraft there at Stallings for training. He then began his introduction to learn to fly the T-33 Jet aircraft at Webb AFB, Big Springs, Texas, where he graduated and received his silver wings as a USAF Pilot. He was then assigned to 18th TAC Recon Squadron, Shaw AFB, Sumter, SC flying the Thunderflash, the RF-84F. While there he completed all requirements to check-out in the RF-101 Voodoo. However the USAF began a Reduction In Force (RIF), limiting check-outs to the rank of Captain and above. He was then a 1st Lt. While limited to minimum flight time due to Air Force budget restraints, he waited approximately 6 months for policy to change. He took an "early out" offer from the Air Force and returned to work at Lockheed.

He returned to his former job at Lockheed (now Lockheed Martin), Marietta, GA, in April 1958 and retired June 1990.While at Lockheed Garland progressed through various assignments and retired as a Staff Engineer. He participated in programs within C-130 Project Engineering and special programs as:

C-130A Production Ski design

C-130BSki redesign for increased gross weight

C-130B, E & H Avionics & Flight Station design and layout of cockpit instrumentation flight crew locations and installation of Avionics Equipment

C-130E Outer Wing redesign

B-70 Valkerie supersonic bomber.

Subcontract portion of fuselage L-1011

Elevator control system redesign

TFRIS Terrain Following Radar Instrumentation System (Prototype) housed in an external drop tank to record flight path, time and spatial parameters for aircraft terrain following mission evaluation.

He was involved in FAA Certification effort of the Commercial FAA Certificated version of the C-130 Hercules which became the Model 382 Series, sometimes referred to as the L100.

There were also two stretch versions, the -20 and -30.

Participated in Certification of the commercial Hercules in Australian Coordinated Engineering response and resolution to Commercial and Ònon-US military customer engineering problems on Hercules series aircraft within the U.S. and World-Wide(approx 15 years)

Coordinated Engineering initiation and review of Service Bulletins for Commercial and "non-US military" Hercules series aircraft (approx 15 years)

Coordinated the annual Commercial Operators Meeting (approx 10 years).

Retired June 1990.

COL JOHN L. LEAPHART

John was born in North, SC, October 13, 1933 and graduated from North High School in 1951, entering Clemson in the fall of that year. When he was growing up he was very interested in becoming a military pilot as he witnessed many of the WWII aircraft that flew out of U S Army's Air Corps North Air Base . So when he entered Clemson he participated in the AFROTC program, graduating in June 1955 with BS Civil Engineering and was commissioned a 2nd Lt USAFR During 1954 he attended AFROTC summer camp at Randolph AFB, Texas where he had first jet flight. From June 1955 to September 1955 he was employed at Lockheed in Marietta, GA as associate aircraft engineer. He then entered active duty during October 1955 and began his preflight training at Lackland AFB, Texas. In December he reported to his basic flight training Class 57E at Stallings AFB, NC and this class at Stallings was the last class to train in the North American AT-6G. He then was assigned to Laughlin AFB, Texas for training in the Lockheed T-33 Jet aircraft. After graduating and receiving his silver wings he was then assigned to Shaw AFB for advanced training in the Republic RF-84F later flying the McDonald FR101 Voodoo. He was stationed during 1958 thru 1960 at Phalsburg AB France flying the RF-101 with winter training in Morocco and Spain. During 1960 thru 1961 he was stationed at Toul AB. France flying the RF-101.During 1961 thru 1964 at Shaw AFB he continued to fly the RF-101, with several European and Far East deployments. He was assigned temporary duty to McDill AFB, FL in 1962 during Cuban Missile Crisis and made several low level reconnaissance missions to photograph various missile sites, etc. in Cuba. During 1964 thru 1965 he was stationed at Laon, AB, France. and worked in Wing Command Post still flying the RF-101. Then in 1965 he was rotated back to the states for temporary duty to transition into the RF-4C. During 1965 thru 1966 he was at Toul, AB, France flying the RF-4C as Chief of Standardization and Evaluation until President De Galle would not allow a continued American military presence in France. During 1966 thru 1969 he was stationed at Mt. Home AFB, ID flying the RF-4C in Standardization and Evaluation. He then was assigned for our months of TDY to Vietnam and Thailand in 1967 flying enough missions to count as a tour. He did a second tour 1969 thru 1970 in Thailand for a South Eastern Asia tour flying RF-4C. During 1970 thru 1972 he returned to Shaw AFB assigned to the 9th AF Headquarters again in Stat Eval flying the RF-101 and the RF-4C. He now had flown about 1400 hours in RF-101 and close to 2000 hrs in RF-4C. During 1972 thru 1973 he attended and was graduated from the Air War College at Maxwell AFB, Montgomery, Al and during this period he also completed his Masters Degree at Troy State University, Troy, Al. in Counseling. His next assignment during 1973 thru 1974 was Battle Staff on the National Emergency Airborne Command Post (Part of JCS in Pentagon) at Andrews AFB, MD. During 1974 thru 1975 he was a Research Associate at the Stanford University Strategic Studies Center in Washington, DC. In October 1975 he decided to retire from the Air Force as a full Colonel with 20 years of active service. During October 1975 thru 1991 he was a stock broker with Merrill Lynch in Washington, DC and later in Charleston, SC. Later he purchased property in Mt. Pleasant, SC in 1979 and retired from Merrill in 1991. After this retirement John continued his career in real estate ventures and personal stock training. He continued to fly Cessna 152's and 172's off and on until about 6 years ago at which time his wife had medical problems and he stopped flying to help care for her during her period of recovery.

LT COL ROBERT J. GOEBEL

WWII Mustang Fighter Ace

Goebel is credited with shooting down 11 enemy aircraft during WWII.

He flew 61 combat missions while a member of the 31st Fighter group, 308 Fighter Squadron. He later earned his College degree in Physics at the University of Wisconsin thru the GI Bill. He flew full time with the Wisconsin Air National Guard and returned to active duty just before the Korean War. He retired from active duty in 1966 as a Lt Col and began working in the NASA manned space program, the Gemini Space missions. He later worked for McDonnel-Douglas in the Skylab Program., and later he held other jobs with Northrop Corporation and Aerojet Ordnance.

Bob Goebel's association with Clemson's Aviation Heritage is thru his brother Norbert B. Goebel. Norbert was involved in running the Forestry Experiment Station for many years there at Clemson and also as a side light, kept the football statistics in the press box for many years. He is buried in the small cemetery on the campus behind the Death Valley football stadium. Bob stated that he went thru flying school with a number of young men from Clemson that were ROTC students but entered flying school before they were graduated as did many students during the war. One person, Claude Greene, who graduated from Clemson Class 41, went thru flying school with him and flew in the same squadron in Panama and later flew combat with him in the same squadron. Borwne Forbes also was a Clemson student prior to going to flying school with him. He later flew transport aircraft in the Pacific Theater.

DR. JIM LONG, PhD

Clemson Professor, Aero Club Faculty Advisor and Private Pilot

Jim was born in Central, S.C., a son of the late W. Homer and Ollie Mae Simpson Long. He retired after 42 years as a Professor of Electrical and Computer Engineering at Clemson University. A graduate of Clemson University, Class of 1943, he earned his masters and doctorate degrees from Georgia Tech.

He served in the U.S. Navy during World War II with the combat Air Services in the Philippines and retired as a Commander from the U. S. Naval Reserve.

Jim was a life member of the Institute of Electrical and Electronic Engineers and continued to work with the National Counsel of Engineering Examinations over the years. He was also a member of the third Clemson Aero Club and Faculty Advisor for that club in the 50's. He was member of the Clemson Rotary Club and Clemson Fellowship Club.

Jim died March 25, 2006 at the Oconee Memorial Hospital in Seneca, SC.

ST JOHN COURTENAY

NASA Marshall Space Flight Center Electrical Engineer

St John Courtenay was graduated from Greenville High School in 1943 and soloed a Piper Cub at the Paris Mountain airport before he entered Clemson to work on a degree in Electrical Engineering. He completed his Freshman Year but he enlisted with a number of friends in the US Army. After serving 18 months in the US Army he was released from service and came back to Clemson to finish the requirements for his degree in Electrical Engineering (EE) and to get a reserve commission in the Army Air Corps. He was allowed to enter the AFROTC Program because of his prior military service. He graduated in June 1951 with his degree in EE and was comissioned a 2nd Lt USAFR. After about two months after graduation he was called to active duty and was stationed first at Lackland AFB, Texas and later he was assigned to Keesler AFB, MS to enter the Electronics school. He was there for about a year and was then assigned to Fuchu, Japan. He was the Director of Radar Intelligence and was sent to Japan. While there he developed a special RADAR simulator for use in Radar Bombing Training. He served for 2 years and then he was released from active duty to return to civilian life. After his release he was employed for a short while with J P Stevens Company in Greenville, SC and late worked for Chemstrand Corp at Decatur, Al. for about 6 years. He next began his employment with NASA Marshall Space Flight Center on October 29, 1962 where as an electrical engineer from 1962 thru 1999 he designed hardware for the Saturn Apollo Saturn Program, Skylab Program, the Space Shuttle, and The international Space Station. He received a number of NASA Awards throughout his NASA career and retired from NASA MSFC after 41 years distinguished of service.

LT JAMES W. WOODFIN

James W "Jim" Woodfin entered Clemson in 1947 to study toward a BS Degree in Agricultural Engineering. He was a member of the Track Team during his Junior and Senior years, active in the AFROTC, a Distinguished Military Student ,and was a member of the Arnold Air Society. He was also a member of the 1950-1951 Clemson Aero Club. He was commissioned as a 2nd Lt USAFR upon graduation, called to active duty, reporting to Bainridge, Ga , for USAF Pilot Training Class 52G. He was later transfered to Bryan AFB to continue his primary pilot training but he and the other student officers were then assigned to Pinecastle AFB, Fl for advanced training in jets where he was graduated and awarded his silver pilot wings. He was next assigned to Manston AFB, England, home of the 406th Fighter Interceptor Wing, and served with the 514th FIS where he flew F-86D's for a few years overseas remaining there till he was rotated back to the states when the 514th was decativated. He was released from active duty and was employed with the US Department of Agriculture as a Hydraulics Engineer in Florida for a number of years. He returned to SC and became a member of the US Coast Guard Reserve retiring as a Lt Commander.

Jim died 19 July, 2005.

COL MARION "GRAHAM" PRITCHARD, JR.

Marion G. Pritchard Jr graduated from Clemson University in 1968 with a BS Industrial Managment and was commissioned a 2nd Lt thru the AFROTC. He was the AFROTC Commander of the Flight Instructor Program that was conducted at the Clemson Oconee Airport by civilian flight instructors and funded by the USAF during 1968. Each participant was required to pass the pilot portion of the Air Force Qualification Test with a certain grade and also a flight physical for regular USAF pilots. Six Squadrons of Cadets participated in the program and 25 cadets obtained ther Private Pilots license as a result of their training.

Marion "Graham" Pritchard, Jr. has had a long career in both military and civilian aviation. He is a former Commander of the 109th Airlift Wing of the New York Air National Guard where he spent over 30 years flying ski equipped C-130s in the North and South Polar Regions. As a civilian Flight Instructor he has taught in single and multi-engine aircraft and seaplanes and has extensive experience in tailwheel instruction.

EDUCATION

B.S Industrial Management Clemson University, 1968
US Air Force Undergraduate Pilot Training, Laredo Air Force Base TX
Air Force Safety and Mishap Investigation Course, Norton AFB CA
Air Command and Staff College, Maxwell AFB
Air War College, Maxwell AFB
Flight Training for civilian certificates and ratings

EMPLOYMENT HISTORY

President, General Aviation Consulting, Inc. 2000 – present
New York Air National Guard, 109th Airlift Wing, 1977 – 2000 (retired)
Federal Aviation Administration, Air Traffic Controller, Syracuse Tower, 1975 – 1977
West Virginia Air National Guard, 167 TAG, C-130 Pilot, 1974 – 1975
Singer Simulation Products Division, Silver Springs MD, Training Specialist, 1974 – 1975
Professional Flight Service, PG Airpark, Washington DC, Instructor pilot, 1973 – 1974
Active Duty Air Force, 1968 – 1973

CIVILIAN PILOT RATINGS

Certificate number 1922402, total flight time over 11,000 hours
Airline Transport Pilot, Airplane multiengine land, type ratings BE-300, BE1900, L-382
Commercial Privileges, Airplane single engine land and sea
Flight Instructor certificate number 1922402CFI, airplane single and multiengine; Instrument airplane

COMMUNITY ACTIVITIES

Member Calvary Church in Charlotte NC
Member Sun City Carolina Lakes Bible Study Club
Volunteer in the Aviation Department, JAARS in Waxhaw NC

RICHARD LEIBY

Mechanical Engineer, Rocket DevelopmentEngineer, and Business Development

Richard was born in Warren, Ohio and later was graduated from Bloomington High School, IL in 1945. His family moved to Sumter, SC in May 1945 and in September he entered (then) Clemson A & M College and was graduated with a BSME in 1949 and also received his USAF 2nd Lt commission thru the AFROTC. At Clemson he played his alto sax with the Brigadiers (Dance Band) for 4 years and also was a Senior Drill Team member. Prior to his entry on active duty he began working for Westinghouse Electric Corp, staying there till 1955, when he was called to active duty with the Air Force. His first assignment was at the Equipment Laboratory at Wright Air Development Center, Wright-Patterson AFB, OH.

Richard was assigned as Project Engineer for an auxiliary power unit (APU) for the KC-97 tanker. The idea was that the airplane would sit in the far north, the APU providing power for crew living. When the call came to refuel B-50s, they were to use the APU to start the engines, shove it out a hatch and take off to rendezvous with the B-50s. (Really.) He later worked on then-unconventional power supplies: solar cells, fuel cells, thermionic and thermoelectric systems. Apparently, there was a thermoelectric gap, and we were behind. The Russians had a thermoelectric generator on tractors and a thermoelectric generator fitted to a kerosene lamp to power a radio! Richard originated a thermoelectric research program that took his successor two years to get cancelled.

After an Air Force-sponsored MSME from the University of Michigan, he wound up in 1959 at the Air Force Rocket Propulsion Laboratory, Edwards AFB, CA as one of two project engineers for the rocket engine on the X-15. (The other engineer, a captain, came to the Lab from a Super C-47, which had a very large fuel tank in the cargo compartment. His job was also to sit in the Far North and await a call. When it came, Richard was to crank up his Gooney Bird and head for Siberia to pick up crew members of B-50s that didn't make it all the way out. They had a static test stand at the edge of the dry lake, essentially the X-15 propulsion system, for trouble shooting and maintenance acceptance. They completed the development test program and then supported the flight program for the first few flights.

Richard went back up to the Lab to an Ultra Low Chamber Pressure satellite system, attempting to field a light weight radiation-cooled propulsion system. He did some interesting things with light-gauge molybdenum, but the program did not work out. Later, another captain and he worked on a propulsions system for a satellite interceptor - although in 1963 they couldn't use "intercept" so they "rendezvoused" instead. The mission analysis was a ball: they thought it would make a great board game. (1963, remember?)

Richard was at the Lab for five years, that last one as chief, small engine section. The hot program was in-house testing of components for ablative-cooled, fluorine/hydrazine engines. Now, fluorine will combust most anything - it has a respectable specific impulse with water as fuel and will ignite sand. Richard's group proved that when, as best they could determine, they got a bit of sand in the fluorine line. Melted almost all of the copper in the test stand. Weapon system offices were not interested in fluorine.

His next assignment was in the Titan III Launch Vehicle program office as engineering officer for the vehicle transfer stage engine. It was Richard's favorite job in the Air Force, involving engineering reviews, change notice evaluation, proposal negotiations, hardware acceptance, flight support and post-flight review. Most of these could be going on simultaneously. As a young captain/major (Richard was there five years, too) he enjoyed juggling five things at once. (Richard says, "Today it's all I can do to juggle one thing!")

Back to Wright-Patterson in 1969 to the AeroPropulsion Lab, Aerospace Power Division. (Richard tells folks that he wasn't in the Air Force, he was in Systems Command.)

Here his branch had a couple challenging programs. Underway when he arrived was a project with Ling-Temco-Vought (LTV) to develop a solid state aircraft electrical system, replacing rotating components and mechanical relays and contactors. They needed a very large (in 1970) power transister, and so they went to RCA.

"How many do you need?"

"Maybe 10."

They weren't laughed out of the plant, but were advised that RCA, who was making millions of these for the automobile "no seat belt, no engine start" systems, was really not interested in an order for 10. A couple phone calls got them the transistors.

The next job was development with Westinghouse to scale up a 120 kva aircraft generator to 4 megawatts (how's that for a "scale up") for an airborne electric laser. (The laser was being developed in another branch of the Lab.) The power supply involved the turbopump turbine from the Saturn V's J-2 rocket engine, a gearbox and the Westinghouse 4mw generator. Unfortunately, the program was not continued.

Richard worked the next four years at Systems Command Headquarters with the Short Range Attack Missile and the Maverick.

He then went back to work at the Rocket Lab in the solid rocket business. As Deputy Director, Richard really enjoyed that work. After the HQ, it was a relief to get back to a real job! The fun here was the interplay with the junior officers-becoming stodgy was not an option.

Richard retired in 1979 from the USAFR as Lt Col and signed on with Rocketdyne Advanced Programs, in Business Development. (Euphemism for Sales.) He retired again in 1988 and then took a half-time job with General Dynamics to follow the NASA Marshall Space Flight Center's engine development program for the Air Force Advance Launch System. That program died after three years and Richard retired for the last time and is still living in Huntsville, Al.

RALPH HOOD

Aviation Professional Speaker, Humorist, and Author

Ralph Hood was born in Anderson, SC in 1941, entered Clemson University, and received his BS Degree in Arts and Sciences in 1964. He has lived in Alabama for many years being employed by as an aircraft sales manager and pilot for Montgomery Aviation, Montgomery, Al and Huntsville Aviation, Huntsville, Al. He has owned a number of aircraft and has many hours of business and priviate flying.

He flew for so many year in Alabama that he became known unoffically as "Alabama's Flying Missionary."

He came to realize that he had a future as a humorist and public speaker and became a nationally recognized professional speaker, speaking generally on aviaton topics toward the humorous approach. He has taught Aviation Managment courses at Southern Illinois University as well as teaching aviation safety to audiences on a national in level in California, Colorado, North Carolina, and Florida. He is also a spokesman for the Aircraft and Pilots Association(AOPA), speaking to aviation and non aviation audiences. For this he is often refer to as the "Flying Humorist". He is the author of several books, newspaper and magazine articles, has his own Internet Blog, and has traveled the US and abroad in his capacity as a certified professional speaker. Although not a member of one of the Clemson Aero Clubs, aviation has played a very important part in his life.

DR. RONALD I. MILLER, PhD

Physicist, Scientific and Technical Intelligence

Dr Miller received his BS Degree in Mathematics and Physics from Austin Peay State University in 1965 and later his Masters in Physics from Clemson University in 1968 and later a Doctor of Science in Physics from Southeastern Institute of Technology, in Huntsville, Al. He has been a teacher on a part time basis at four southern Universities.

He began working at Boeing in 1969 and continued till March 1977. Dr. Miller performed theoretical studies and developed experiments for NASA research programs during 1972 thru 1977 for the solidification of materials in the micro-gravity environment of space. His research work from 1969 thru 1972 also supported the Apollo Program, the Skylab, the Apollo-Soyuz mission, and the Shuttle, and he conducted various theoretical studies supporting experiments for sub-orbital rocket flights. These studies laid the foundation for materials-science-in-space projects flying on the Space Station today.

At the Defense Intelligence Agency's Missile Space Intelligence Center (MSIC) from 1977 thru 2009, Dr. Miller was responsible for the management and technical direction of scientific and technical analyses of foreign directed energy weapon (DEW) systems. These systems include laser, radiofrequency, and particle beam weapons, as well as electro dynamically launched, hypervelocity kinetic energy weapons for the anti-satellite, anti-ballistic missile, and air defense missions. Intelligence analyses include developing mathematical models and computer simulations of these systems, and performing engagement simulations of the DEW systems against U.S. targets to assess the DEW threat to U.S. forces and interests. While serving as Chairman of the Intelligence Community's DEW Subcommittee from 1990 to 1998, Dr. Miller managed these efforts for all DEW missions at the national level. In both his MSIC and Committee positions, he served as advisor to the Departments of Defense, State, and Energy, and the Congress on foreign DEW systems and technology.

Dr. Miller is the author of more than sixty scientific journal articles and government reports in the fields of liquid state physics, low temperature physics, electromagnetic field theory, and laser science/systems engineering. He is a member of the American Physical Society, the Directed Energy Professional Society, an Associate Fellow of the American Institute of Aeronautics and Astronautics, and an Elder in the Presbyterian Church. He has received numerous NASA, Intelligence Community and DoD honors, in particular the National Intelligence Medal, which he was awarded in 1999. Dr. Miller is listed in several biographical reference works, including "Who's Who in Aviation and Aerospace" and "Who's Who in America" and was inducted into the Phi Kappa Phi Academic Hall of Fame at Austin Peay State University in 2004. He and his wife, Jan, have two grown daughters, and two grandsons.

WARREN JEFFORDS

Warren Jeffords graduated from Clemson BS Engineering in 1949. He began working for Chrysler during the Ballistic Missile programs in Huntsville that began in the late 50's as a manager for Chrysler. During this time Chrysler was the Army's research and development contractor for producing the Redstone and Jupiter Missiles as well as supporting the Army Ballistic Missile Agency's (ABMA)Missile Firing Laboratory at the Kennedy Missile Test Range. Warren was known to many of the ABMA civil servants that he worked with as "Mr Chrysler".

HEYWARD B. DOUGLAS

Originally from Columbia, SC, Heyward attended Clemson University and received a B.S. in 1969 and M.S. in the Biological Sciences, graduating in 1971. He served as a pilot/ Instructor Pilot in the USAF from 1971 to 1976. Douglas returned to Clemson University in 1976 and worked as an Extension Entomologist (Insects) until 1979. From 1979 to 1981 he earned his living as an aviator again by towing banners, dropping skydivers, taking aerial photos, flying charter flights, and giving lots of flight instruction. In 1981, he returned to Clemson Univ. again, this time as the administrative pilot, flying administrators, faculty, and staff (and sometimes coaches) all over the country.

He is married to Carolyn Douglas and they have 3 grown children. Working with Boy Scouts has been a primary avocation for many years, and as part of that, Heyward has completed the Woodbadge leadership training. He began hiking on the Foothills Trail in the early 80's, and in 1989 joined the Board of FTC, where he has served as chairman, maintenance coordinator, and as a shuttle driver.

Heyward is still flying as a pilot for Clemson University.

CONCLUDING REMARKS

Researching and developing this manuscript has been a "Love of My Life." When the time came to stop researching and start writing the book it was difficult to stop but somewhere one must stop researching and develop a final manuscript for publication. Although I was a day student and hitch-hiked from Seneca to Clemson each day I participated in many of the daily Cadet military functions. Yes, I was a Rat and was treated as such but still the interaction with students and the importance of the Corp molded me.

The purpose in creating this book was to show my appreciation for the Corp and Clemson's Education System that allowed me to be successful in my later life endeavors. Clemson University has had an outstanding number of Military Graduates who have brought credit to Clemson Military so I felt that some of that history should also be documented in this book.

So to the former Graduates, those to graduate this year, and the future Graduates I say "AimHigh, The Best is Yet to Come."

Otha H "Skeet" Vaughan Jr
Huntsville, AL
2011

APPENDICES

APPENDIX 1

Some Clemson Flying Club Instructors

James Jackson.....Chief Instructor for 1948 Clemson Aero Club at the Clemson-Central Airport
D. Mike Smith....During the 60s he instructed Aero Club members at the Cherry Bottom Airstrip
Reid Garrison.....During the 60's he flew out of the Cherry Bottom Air Strip and the Anderson Airport while working for Ben Johnson's Flying Service. He was the FBO at the Clemson-Oconee Airport 60's and early 70's.

Marshall Robertson.....During the 60's at Anderson Airport he instructed a number of Aero Club students while at the Anderson Airport.

Kevin Prufers.....During 1977-1979

Kevin, after graduating from high school in NC, came to Clemson in 1977 to work toward a degree in Economics and received his degree in 1979. Before he came to Clemson he already had his private pilots license and he joined the Aero Club. By 1978 he had obtained his Commercial Flight Instructor license, his Instrument rating, and Multi-engine ratings. He instructed for Reid Garrison at the Anderson Airport and instructed for the Aero Club as well. Some of the students he instructed during this time were:
David Sechrist, Nelson Graham, Chip Houck, Phil Robertson, Kevin Carey, Cris Foreman, Gene Rochester, Art Chaknes, Tim Hall, Eddie West, Kirk Crawford, Steve Simpson, Jeff Reynolds, Dan Rankin, Ken Brown, Pete Dimopoulos, Paul Barnett, Bob Turner, and Margaret Hancock. After leaving Clemson he went into commercial aviation and then later was hired by Piedmont Airlines as co-pilot during 1985, advancing later in 1985 to Captain. He still continues to fly for Piedmont Airlines, now USAir.

Dr. Richard Figliola.....During the 80's and 90's at Clemson Oconee Airport
George Rowland.....During the 80's and 90's at Clemson Oconee Airport
John Boobas.....During the 80's at the Clemson Airport.
Brandon Arrington.....2006 at Oconee Regional Airport
Robert Banks.....2005 Airport Manager and Club Flight Instructor
Pat Kohout.....Flight Instructor 2003 thru 2005
Terry Leap:.....Flight Instructor and has been a Faculty Advisor 1983- Present

He is Chair of the Management Department in the Clemson College of Business and Public Affairs. He is a native of Haninabal, Mo and has taught management at Clemson University since 1983. He has been affiliated with CUFC since 1983, and he has served as a flight instructor with the club since 1991. He learned to fly as a teenager and soloed at the age of 17. He has single- and multi-engine commercial ratings as well as a Jet Aircraft Citation type rating. He provides both primary and advanced instruction. His success stories include Jon Pasek (owner and manager of Greenville Aviation Flight School) and Jeff Treece (Air Tran and US Air), and Justin Sawyer (Comair).

Shane Valentine.....Part time Instructor
Jim Anderson.....Part time Instructor
Randy Collins.....1989-Present

He is an Associate Professor of Electrical and Computer Engineering at Clemson University, joining the Department in 1989. He is an active rated Airline Transport Pilot and has been a flight instructor for a number of years for the flying club and has been a Faculty Advisor

Jason Williams.....Part time Instructor
Heyward Douglas.....Faculty Advisor and Flight Instructor
Tony Barr.....Flight Instructor 78-79 time frame

APPENDIX 2

Some Faculty Advisors for the Clemson Flying Club over the Years

Professor John Marshal.....1927 thru 1941
Major C D Foster 1949
Joe Rostron .....The 50's
Dr. Jim Long .....The 50' to 60's
Dr Alan Cover..... The 60's
Dr E.P. Stillwell .....The 60's
Dr Don Turk .....The 60's thru the 80's
Dr. B. J. Todd .....The late70's and Early 80's
B. J. Boykin .....The 80's to the 90's
Heyward Douglas.....The 80's to Present

APPENDIX 3

Aircraft that Clemson Flying Clubs have owned from 1948 till Present

1. Piper J-3 N4128 First plane purchased by the Club
2. Piper J-3 N41026 Second Plane purchased
3. Taylorcraft BCD-12, N39107 Bought in 1950 sold in 1953
4. Funk B, 1946 Model, N81190 Crashed at Cherry Bottom Field, sold, Was flying in 1990 but now not registered.
5. Aeronca D7, N49546, Bought in the 50's, sold later to a museum, sold and later crashed at Clemson. Laugermilk's son was pilot. He was killed and a passenger injured while flying over Lake Hartwell near Clemson and hit power line.
6. Piper PA-22-08, N4943Z, Bought in 1961, sold, Now was registered in Maryland.
7. Taylorcaft BC-12D, N95809, still flying in Alabama.
8. Cessna C-150, N8668C.
9. Grumman AA-5-0526 Tiger, N9526L,Gift to the Club by Dallas Sherman, crashed, sold, and now still flying.
10. Cessna 152, N65407, still flying.
11. Cessna 172, N737RP, Damaged by another Aircraft, rebuilt, still flying and owned by Randolph Air Service of Owosso, Michigan.
12. Cessna 172, N733VS, Bought 2003.
13. Cessna 150, N6205G, Was in Club prior to 1981 Was sold and now in Casper, Wyoming and owned by Robert N. McDaniel.

APPENDIX 4

Flight Indoctrination Program (FIP)

Prior to 1991, AFROTC also conducted a Flight Instruction Program (FIP) parallel to the Pilot Indoctrination Program (PIP) at USAFA. Although often touted as a means for AFROTC cadets to earn a free FAA Private Pilot Certificate while in college, the actual intent of the program was to provide an additional flight training screening process for prospective USAF pilot candidates. In AFROTC, FIP consisted of two blocks, the first being a private pilot ground school course taught by an aeronautically rated USAF officer assigned to the AFROTC detachment's cadre. The ground school course was also given an AS400 series designation and open to all AFROTC cadets in their senior year regardless of selection or non-selection for USAF undergraduate pilot training. Cadets who had prior civilian flight training and/or civilian pilot certifications could also enroll in the FIP ground school and the course was also offered as an option for Army ROTC cadets, Naval ROTC (NROTC) midshipmen on both Navy and Marine Corps commissioning tracks, Naval Aviation Reserve Officer Candidates (AVROC) and Marine Corps Platoon Leaders Class-Air (Marine PLC-Air) officer candidates slated for flight training in their respective services following graduation.

The flying portion of FIP was typically conducted by civilian instructors under USAF contract at a nearby civilian airport, normally employing light general aviation aircraft such as the Cessna 150 / Cessna 152 series, Cessna 172, Piper Cherokee, or other similar aircraft. Since FIP was designed as a washout/attrition device, AFROTC cadets who already held an FAA Private Pilot's Certificate or greater were not eligible for any actual flight time via FIP. Those cadets without prior flight experience initially received 38 flight hours, but post-Vietnam War defense cutbacks in the mid-1970s resulted in FIP being reduced to a "safe for solo" program with 25 hours of funded flight time. FIP was discontinued in 1991 when it was replaced by the single-site Enhanced Flight Screening Program (EFSP) at Hondo, Texas.

APPENDIX 5

Some Former Professors of Aerospace Studies at Clemson's AFROTC Detachment 770

1951-1952.....Col John B. F. Dice
1952-1958.....Col Lloyd H. Tull
1958-1962.....Col C. B. Thompson
1962-1963.....Lt. Col J. D. Cotter
1963-1964.....Lt Col G. H. Wilson
1964-1967.....Lt Col Dewey R. Franklin
1967-1972.....Col Elmer N. Tyndall
1972-1975.....Col Duncan W. Rabey, Jr.
1975-1978.....Col Charles R. Lakins
1978-1981.....Col Edwin F. Rumsey
1981-1985.....Col Lewis E. Jordan
1985-1988.....Col William R. Austin II
1988-1989.....Col Grover E. Musselwhite
1990-1992.....Col Tyler B. Huneycutt III
1992-1994.....Col Alton C. Whitley, Jr.
1994-1997.....Col Marc L. Drinkhahn
1997-2001.....Col Sandy Edge
2001-2004.....Col Edward B. De Iulio
2004-2007.....Col Lance S. Young

AFROTC Detachment 770 Notes

Detachment 770 has proven itself year after year as one of the nation's top medium-size AFROTC detachments in both the Southeast region and in South Carolina as noted by several prestigious awards: Right of Line, High Flight and Best in State. Furthermore, Detachment 770 is nationally recognized for its dedication to generating quality leaders for the U.S. Air Force. On average we commission 16-20 second lieutenants, and more than 80 percent who apply for rated (pilot, Navigator) receive rated positions. Clemson AFROTC has a very rich military heritage dating back to WW1.

With the National Security Act of 1947 establishing the Department of the Air Force, Clemson's Air Force ROTC Detachment was first created, beginning a long tradition of producing the finest officers for the world's greatest Air Force. In 1955, concurrent with the acceptance of women into Clemson College, Clemson's Cadet Corps was abolished and Clemson became a civilian school. Although the Cadet Corps was abolished, participation in at least the first two years of ROTC was still compulsory for all male students. In 1964, in recognition of expanded offerings of the institution, the name of the institution was changed to Clemson University. By 1968, the ROTC requirement for underclassmen was abolished, stemming a decrease in Detachment 770's size. In 1979, the detachment was reorganized. Today, it is subordinate to the Air Force ROTC Southeast Region. Air Force ROTC is an integral part of campus life at Clemson University, stemming from Clemson's early history as a land-grant, military college. Clemson University has a proud military tradition that dates back 117 years, with more than 10,000 Clemson alumni who have served in the Armed Forces. To date, 473 alumni have been killed from WWI through the current wars in Afghanistan and Iraq.

As of 2001, AFROTC Professors of Aerospace Sciences are required to have a BS degree and a Masters degree. They should be first-line officers, Lt Col or Col, but they do not have to be rated. They must be approved by the University where they are to be in charge of the AFROTC program. They hold a faculty postion, usually in the School of Business Administration.

Students cannot major in Aerospace Sciences at Clemson as it is an elective minor.

DET 770 Awards

1984.....Line of Right Award
2005-2007.....Arnold Air Society National Headquarters located at Clemson University
2005-2007.....Air Force Association Best Detachment in South Carolina

APPENDIX 6

Some Oconee Clemson Airport Managers

Reid Garrison
Jim Thomason
Marion Lyles
Kevin Short
Alan Laudermilk
Ray Smith
Robert Banks

APPENDIX 7

Arnold Air Society, Silver Wings, Senior Platoon, and Pershing Rifles Notes

William H Fanning
Ronald H. Clayton
James A Woodfin
Franklin D. Burdette
John D. Skarrett
Edward J. Hildebrand Jr.
Alan B. Killingsworth
John C. Wilson

John J. James Jr — Clyle K. Grambrell Jr
H.D. Dickbert Jr — John J. Weeks

These people above signed the charter in November, 1951

The 1950 TAPS picture for the Arnold Air Society shows the following members of the first Arnold Air Society as follows:

Herbert O Abercrombia — W. M. Barfield
J. P. Bland — R. H. Bouchard
J. E. Bumley — Eugene C. Carter DMS 49
Julian F. Craig Jr DMS 49
John W. Deas DMS 49 Secretary and Treasurer
D. Peake
Roy K.Frick DMS 49 Squadron Commander
Hollis B. Hance Jr DMS 49
M Herbert — R. T. Hewitt
R E. Hiller — J. I. Howell
A. B. Killingsworth — J. F. Lanham
L. A. Mitchel — John E. Mostella DMS 49
B. J. Mullin — E. A. Smith
William F. Taylor DMS 49 — J. P. Uldrich
Carol H. Warner DMS 49
Mac A. White Jr. DMS 49
John C. Wilson — James L. Whitlaw DMS 49 Adj. and Recorder
B. J. Woodham DMS 49 Executive Officer
St John Courtney Not in individual pictures but in the group meeting picture.
Otha H. Vaughan Jr. DMS 49 was not present when picture was taken.
The following additional members of the AFROTC were as follows:
Albert A. Peters DMS 49 — Charles W. Clements DMS 49
Robert D. Donovan DMS 49 — Olie L. Stokes DMS 49
Frank O. Haslden DMS 49 — Peter W McCravery DMS 49
Alexander H. McLaurin DMS 49 — Thomas M. McCurry DMS 49
J. L. Huckabee DMS 49 — Marion W. Cornwell Jr DMS 49

To be a member of the Arnold Air Society in 1950 you must be an AFROTC student and based on your military class standing you were selected by the AFROTC as Distinguished Military Student (DMS) or for other outstanding qualifications.

The Distinguished Military students were eligible to apply for a direct commission as 2nd Lt in the Regular Air Force just prior to their graduation.

A Brief Review of Clemson's AFROTC Angel Flight History

The first Angel Flight was Founded at the University of Omaha in 1952. Untill 1957 The Angel Flight was primarly a National idea with many names, uniforms, and activities. During April 1957 at the 8th National Conclave of the Arnold Air Society the Angel Flight became a national organization to be unified in purpose and ideals.

The Clemson Flying Tiger Angel flight was organized during the spring semester of 1963. It was sponsored by the William A. Bevan Squadron of the Arnold Air Society there at Clemson. The flight received their charter from National Headquarters on Feb 1, 1963. Major Sarah E. Cochran was the Commander and its Executive Officer was Captain

Bobbie Reed. The flight participated in several campus-wide activities as an immunization program and blood drives. The flight also sponsored several drop-ins and joint socials with the Clemson Arnold Air Society.

At one time the flight membership declined to a minimum level but it was reconstituted through the hard work of Major Dennis H. Sattler and its members. The organization flourished and achieved national recognition by receiving the Ann T. Hawkins Award on both regional and local level. To honor Major Sattler in 1982 the Angel Flight was named the Major Dennis H. Sattler Angle Flight.

During 1984-85 Katherine Ann Valdes became National Commander and while serving in the Angel Flight she became ROTC Cadet Vice Wing Commander. She was graduated in May 1986 and became a 2nd Lt USAFR and entered active service.

During the 1996 National Conclave the national staff proposed a major change in the organization. The members would no longer wear uniforms or have titles as the Arnold Air Society had. The Societies would no longer be called flights, or have rank and ribbons. They would be Chapters, with a President, Secretary, Treasurer, and other such positions.

During 1998 National Conclave the organization voted to change the name to Silver Wings.

During 2004 National Conclave a National Awards System was established for the Chapters and National Awards for Individuals was established.There were to be also Region Awards and Chapter Awards for outstand service to the ideals of the Silver Wings

The Mission of Silver Wings

Silver Wings' predecessor organization, Angel Flight, had a specific objective of supporting the mission of Arnold Air Society. While this is no longer explicitly stated, Silver Wings remains committed to developing strong civilian leaders with an understanding of aerospace power–and obviously, the continuing partnership between AAS and SW is critical to building this understanding.

As specified in the Silver Wings National Constitution, Silver Wings (SW) is a national, co-ed, professional organization dedicated to creating proactive, knowledgeable, and effective civic leaders through community service and education about national defense. This mission includes the following interrelated objectives:

Personal Development of Individual Members:

Local chapter activities and projects provide character growth through service to the community, which enables the individual member to develop a sense of purpose within and a responsibility to the local community.

Professional Development of Individual Members:

Local chapter activities and the national organizational structure provide members with leadership opportunities and the chance to develop professional skills. These include, but are not limited to oral/written communication, public speaking, and project planning and execution.

Civic Awareness:

Exposure to Air Force issues and personnel enables members to develop a personal awareness of aerospace power and the role that it plays in the national defense. The leadership and skills development objectives are enhanced by opportunities for members to actively promote aerospace issues in the community through educational programs, seminars, public relations activities, and voter awareness projects.

While Arnold Air Society and Silver Wings are separate organizations, our strong partnership is based on highly complementary missions. The focus of the Arnold Air Society mission is to build strong officers for the United States Air Force. Silver Wings also focuses on leadership development, and both organizations work together to further each others' goals.

In 1990-1992 the Silver Wings National Headquarters was located at Clemson University and the Headquarters was located again at Clemson during in 1997-1998.

From 1974 when the first woman Kathleen Morris Hager was commissioned thru Clemson's Det 770 AFROTC program there have been a total of 79 women who received their USAF Reserve commissions thru 2006.

History of the Clemson Senior Platoon

In 1928, a Clemson Cadet, F.B. "Gator" Farr '30, had a continuing argument with his brother who was a student at The Citadel as to which school was the best. The Citadel had a fancy drill platoon that traditionally won the summer ROTC drill competition. They decided that the only way to settle the argument was for Clemson to enter the competition. Gator came back to Clemson and organized a drill platoon to enter the competition at the ROTC summer encampment in the summer of 1929. Clemson won the competition.

The drill platoon continued drilling and was called "The Fancy Drill Platoon" or "Junior Platoon." In 1934, the name was changed to "The Senior Platoon" as its members were all seniors or rising seniors.

The drill platoon continued to perform at football games, parades, and special events from 1930 to 1960, creating over 1500 Alumni. There are approximately 600 still living. The Senior Platoon was disbanded after 1960 when the elimination of the Cadet Corps caused a lack of interest from the students.

The Senior Platoon has drilled from Boston to Dallas, to Miami to New Orleans at professional football games, baseball games, the Cotton, Orange, and Sugar Bowls, Mardi Gras and many other activities. At one of the professional baseball games, the Duke and Duchess of Windsor were present in the stands.

In 1992, under the leadership of Leonard Butler '53, Senior Platoon Alumni began having reunions. In 1996, with the ROTC program under probation, the University administration asked if the platoon thought they could field a drill unit and perform the routine at halftime of a football game. On November 16, 1996, "The Clemson Alumni Senior Platoon" performed during halftime of the Clemson-N.C. State football game. The members' age ranged from 60-72. They drilled at halftime of football games in 1997, 1998, and 2000. In April 2002, at a Pass in Review ceremony, the Sabre was passed from the "Alumni Senior Platoon" to "The Tiger Platoon," consisting of students of all classes to represent Clemson University as the Fast Cadence, Precision Drill Platoon.

The Clemson Alumni Senior Platoon has passed these duties on to younger bodies but their love and respect goes to the families and friends of the ones who have made the supreme sacrifice for the freedoms we enjoy.

(From Tiger Platoon Website)

Remembering the Senior Platoon

A Clemson tradition from the past.
Sept. 23, 1998
by Emily Rabon

In 1959, one of the greatest Clemson traditions came to an end. This was the final year of the Senior Platoon, that began in 1930 under the guidance of the legendary Frank B. "Gator" Farr.

The fast stepping (double cadence) drill team performed intricate maneuvers that always ended with the Queen Anne salute. They were always in demand to perform and made appearances at various parades, including the Azalea Parade in Charleston, the Shriners Festival in Aiken in front of Clemson alum and then governor Strom Thurmond.

The platoon also traveled as far away as New Orleans to march in the Mardi Gras parade. Two of the most prestigious places they have performed were Yankee Stadium and Fenway Park. They also performed at all home football games and sometimes traveled with the team.

The only time there was not a Senior Platoon was during World War II because most of the older cadets were in the war. During this time the Tiger Platoon was formed, which was opened to all cadets. The ending of the Senior Platoon came in 1959-60 when Clemson became a civilian institution.

Before the change, the platoon was a highly sought after honor to the cadets. Tryouts were held in the spring and only rising seniors could tryout. Anywhere between 100-150 cadets tried out every spring and fifty were chosen for the forty-three member drill unit. There was competitive training for one month and the old members vote on the new members. If a cadet made a single mistake, then they would not be asked to come back. After making it through the rigorous month of competitive training, the cadets still had to undergo a five minute interview in front of the current platoon. You would be interrogated on the various routines and the potential member had to display a positive attitude.

If the cadet made it this far, there was still more to undergo. They then had to endure two weeks of initiation. They carried around a box of "goodies" and a paddle. Every time they saw a current senior member, they did a Queen Anne salute and give the members candy in exchange for a signature on their paddle. After lunch, the new members were lined up outside of the dining hall and paddled by the old members.

After the rigorous and sometimes humiliating two weeks, the tables were turned and the new members sought their revenge! After their final "paddling", the new members shook the old members hands to signify the switch from old to new. Then, the new members would take the old members and throw them in the reflecting pond on campus. "We looked forward to that for the entire two weeks," remembers former cadet Leonard George, a 1952-53 member of the platoon.

All of the training and torture was all worth it though. If you made it, you were part of the elite. But there was still more to pay. "We drilled from 6 a.m. until 6:45 a.m.," remembers George, "Then we went and got ready for breakfast and classes. At 4 p.m., we were back at it again. But it was worth it. When we competed against other schools, they did not even compare. At half-time of the football games, instead of the fans getting up and going to the concession area, they would stay to watch the platoon perform. It was electrifying to watch."

Former Sports Information Director Bob Bradley remembers the team. "One minute, they would be spread out all over the football field and you would think that there was no way they would get back together. But the next thing you know, they were all back in line and still in step together."

George was the commander of the 1953 Senior Platoon when they traveled to New Orleans for the Mardi Gras parade. "That was the most difficult parade we ever marched in. The crowd was loud and out of control."

In the early 1990's, George was talking to Debbie Dubose, director of Alumni Relations, and he suggested that they try to reunite the members of the Senior Platoon. They ended up contacting over 1500 members and the first reunion was held in the President's Box after the 1992 N.C. State football game. One of the members decided that the best way for people to remember the platoon was if they could see them perform. In the summer of 1996, the members met in Clemson and practiced one weekend each month from June until October. Their debut was at the 1996 N.C. State football game. "We made a few mistakes. The crowd was so loud we could not hear the some of the commands!" remembers George, "Last year when we performed at the Duke game, the announcer asked for the crowd to be quiet until the end. We were perfect that time. I just hope we can pull it off again this year.

We are starting to get up there in age."

What makes these dentists, retired surgeons, retired textile managers, engineers and PhDs get together each year? "It is just the love for Clemson," says George, "When there are members driving all the way from Texas and Virginia once a month to practice, their dedication shows. This group is a wide cross section of Clemson humanity. This group is a good focal point of Clemson spirit."

Notes on The Pershing Rifles

In 1891, General Pershing, then a 2nd Lt in Troop L, 6th Cavalry Regiment at Fort Bayard, New Mexico, became a professor of Military Science and Tactics at the University of Nebraska. Pershing wished to increase the morale and discipline of the battalion there, as well as to increase support for the Cadet Corps throughout the university's staff and community. To this end, he formed a hand-picked company of men, known as Company A, and made them his premier drill unit. This company A eventually promoted the establishment of the national organization known as the Pershing Rifles.

Company Charlie-Four of the National Society of Pershing Rifles was founded at Clemson University in 1939 as an elite drill platoon. Only the best drilled cadets were offered membership and it was considered to be a high honor. Pershing Rifles is the only drill team at Clemson to have survived through the years and it is one of the last few reminders of Clemson's Cadet Corps, military traditions, and military history.

Company C-4 plays a very active role in the univeristy and the surrounding communities by performing military drill and ceremonies which include: drill routines, parades, color guards, military funerals, and military weddings, to name a few.

Pershing Rifles retained their National Championship trophy. The Clemson unit, Company C-4, has won the national chamionship five out of the last seven years as of 2011.

APPENDIX 8

Clemson University Corps Notes

Clemson became a University in 1965 and the ROTC program was not mandatory for the incoming freshmen and sophomores to participate in the ROTC program.The number of freshmen and sophomores were not required to participate in the ROTC programs so the ROTC program was in danger of being eliminated by the ROTC which funded its operation.

A group of Clemson Alumni and retired military Clemson graduates decided that needed to be done, so while attending the 1997 Veterans Day football game at Clemson, the group formed an Association called The Clemson Corps. The group consisted of the following people:

Paul E Blackwell Class 63 Lt Gen US Army Retired (Former President of the Army Alumni Association)
William C Laffoday Major USAF Class 51 Retired
Richard W. Maddox Class 51 Col US ARMY Retired
David Mootrehead Class 51 Col US ARMY Retired
Sandy Edge Class 72 Col Commander of AFROTC Det 770
Ann B Smith Class 81
William Millen Lt Col Commander ARMY ROTC

They proposed the formation of the Clemson Corps, a constituency group dedicated to the preservation of the University's Distinguished Military Heritage. The purpose of the Corps was:

To contine Clemson's Military History
To support Clemson's Army and Air Force ROTC programs thru a scholarship endowment
To develop other opportunities to commemorate the military history of Clemson University.

The Clemson Corps is now called the Clemson Corps Committee and is part of the Clemson Alumni Association, as was its initial intent from the Corps' inception.

Life Emeritus members of the Clemson Corps Committee:
Lt General P. Eugene Blackwell.....Class of 1963
Major William C. Laffoday.....Class of 1951
Col Jack W. Carter.....Class of 1970
Col Danny I Rhodes.....Class of 1968
Brig General C.R. Hap Carr.....Class of 1960
Honorary Directors
George U. Bennet.....Class of 1955
Dr. Robert C. Edwards.....Class of 1933
Dr. Jerome V. Reel.....HA Class of 1999
Col Beverly N. Skardon.....Class of 1938
As of 2006 the following were active Committee members:
Lt Col (retired) Willie Archer.....Class of 1957
Col (retired) William Austin.....Class of 1959
Will Brasington.....Class of 1972
Capt. (retired) Charlie Bussey.....Class of 1957
Marvin G. Carmichael.....Class of 1971
Col (retired) Bernie Chapman.....Class of 1971
Lt Col Frederick Choi
Amy Csernal Craft.....Class of 2004
Col (retired) Edward De Iulio
Col (retired) Sandy Edge.....Class of 1972
Col (retired) Nick Fletcher.....Class of 1969
Earl Gaulden.....Class of 1951
Lt Col (retired) Franklin D. Hartsell.....Class of 1960
Teresa C. Hopkins
Maj (retired) Dawson Luke.....Class of 1956
CMDR (retired) J. David Lyle.....Class of 1968
Lt Col (retired MSC) William F. Mays.....Class of 1957
Brian O'Rourke.....Class of 1983
Capt (retired) Philip Piazza
Ann B. Smith.....Class of 1962
Tom Turner.....Class of 1951
Matthew J. Watkins.....Class of 1977
Lt Col (retired) Ken Weaver.....Class of 2004
Col Lance Young

As of 2007 there were 35 members of The Clemson Corps Committee.

Since 1997 a total of 352 scholarships have been awarded to Army ROTC and Air Force ROTC students through 2007.

Military Plaza

The Military Heritage Plaza is a commemorative square dedicated to the schools's Cadet Corps as an intergral part of the educational process from the school's founding until the mid 1950's. The plaza also recognizes the military contributions of Clemson's alumni from the Spanish-American War through the Persian Gulf War.

The design features two columns that form a three dimensional silhouette that symbolizes how Clemson has molded students into military cadets. Other design elements include a prominant life-sized bronze stature of a cadet, a series of bronze footprints in a platoon formation, words are cased in the entrance steps describing the qualities of cadet life, and there are more than 70 military decorations that are encapsulated in Leucite "bricks" located in a sailor course of walls and columns.

Through the art and the landscape, the plaza seeks to convey factual history as well as deep sentiment. The plaza is located on the north slope of Boman Field adjacent to Tillman Hall.

Major Donors

Ashmore Bros Inc
Class of '41, R Roy Pearce, President
Class of '42, Thomas C Breazeale, President
Class of '58, Clyde W Mulkey, President
Milton W Holcombe and Family '53
Senior Platoon Alumni

Alumni Who Guided the Effort to Erect The Military Plaza as a Tribute to Clemson's Military Histories are listed as follows:

Russell C Ashmore Jr '50
N James Covington '51
Daniel M Carmichael Jr, Class President '53
William W Coogler Jr '53
James E Cushman '51
LeRoy H Doar Jr, Class President '52
F Earle Gaulden Jr '51
Samuel J Grant Jr '52
Bobby L Hughuley '52
William C Laffoday, Project Chairman, Class President '51
John E Lunn '52
Thomas C Mann '51
W Richard Mattox '51
John H Pitts III '51
Ann Batson Smith '82
Thomas P Turner Jr '51
Samuel M Willis, Class President '50
Hughey A Woodle Jr, Fund Campaign Chairman '51

In the Footprints of a Soldier

by Kelly Newman

This article originally appeared in The Tiger on October 22, 2010

For some, it's concrete steps with words inscribed on them. For others, it's a statue often mistaken for a human. But for Clemson military alumni, the Military Heritage Plaza located in front of Tillman Hall is a symbol of the journey from cadet to soldier.

Since its establishment in 1893, Clemson has had a longstanding military tradition initially as a military college and in the present day with its extensive ROTC program. Military Heritage Plaza is a memorial to that exemplary tradition and honors the Clemson alumni who proudly served the United States during times of war. Built in 1996, it was a gift from the classes of 1950, 1951, 1952 and 1953.

The plaza is comprised of four levels, each of which represents a year as a cadet. The steps connecting the levels are engraved with different qualities or traits each cadet is expected to learn and live by after each year in ROTC. For example, the first set of steps includes Discipline, Teamwork, Confidence, Humility, Equality and Courage, which represent the qualities cadets should have after their first year. The other steps include Respect, Responsibility, Perseverance, Honor, Pride, Esprit de Corps, Leadership, Integrity, Success, Loyalty, Life and Duty.

On the second level of the plaza, there are a series of shoeprints of brave men who served their country. Each set is unique to a Clemson alumnus who served in combat. One significant example is Col. Sandy Edge, whose footprint has a tiger paw imprint in the heel, as he continues to work for the university to this day. Along the walls on the plaza different medals are displayed, each of which were received by Clemson alumni.

Another significant part of the plaza is the mold of the cadet between the two pillars, half of which is on the inside of each pillar.

This mold is a physical representation of what each cadet will become after his or her four years at Clemson. Likewise, the statue at the base of the plaza is an example to all cadets of the proud military tradition they will represent. "For me, it's an inspiration to carry on the tradition of the cadets who came before me," Eric Basso, a sophomore in Army ROTC, said.

Over the past 14 years, the Military Heritage Plaza has become an integral part of the ROTC tradition at Clemson. "I have a great deal of respect for Military Heritage Plaza and those former cadets of Clemson for whom it was built," Cadet First Lt. Sam Davis of the Air Force ROTC said. "It reminds me that I am following in the footsteps of countless Tigers that have served their country proudly, and it pushes me to do the same." Cadets treat the plaza with the utmost respect and have their own customs when on it.

For example, cadets don't step on the plaque in front of the plaza or the various footprints on the second level. Also, cadets don't walk down the center of the plaza over the listed qualities on the steps or through the mold.

At their Commissioning Ceremony, these displays of respect become even more significant for the cadets as they finally proceed one by one down the center of the plaza to receive their first salute and give a silver dollar to the enlisted person from which they got that salute as a sign of mutual respect.

The rich tradition Military Heritage Plaza represents goes beyond the ROTC program on campus; it honors not only the military legacy on campus but also the Clemson alumni who so valiantly fought for our country and made unfathomable sacrifices, which is something that everyone can respect and admire.

Current Issue:
April 22, 2011
Volume 105 | Issue 12

Future Projects
Scroll of Honor

The origins of the Scroll of Honor can be traced to 2002, when Col Danny Rhodes (Retired) (Class of 1968) and Dawson Luke (Class of 1956), a retired Army major who twice served active duty in Vietnam, began collecting names of Clemson alumni killed in conflicts from World War I to the current wars in Iraq and Afghanistan.

In 2005, BG Hap Carr, Rhodes, Luke and many others with the Clemson Corps began working on the concept. They received approval from the university's administrative council in March 2007 and began fundraising that June. Since then, the committee has worked with architectural firms and sculptors to formalize the concept, and the construction began in 2006 with completion and dedication planned for 2010.

APPENDIX 9

Missile and Space People

These are some of the people who worked in the Army Missile Development programs and later in unmanned space flight missions and some later were involved in the NASA manned Apollo Lunar Landing Program. These people were as follows:

James Sloan, Jim Robinson, Max Strauss, Martin Geiger, St. John Courtney, William Buist, J. J. McLaughlin, Lt Henry McGill (Aid to General John B. Medaris, Commander, Army Ballistic Missile Agency,), Gerald Arnts, Warren Jeffords, Bob Abbot, Terry Vaughan, George Norris, Joe Sipple, Roy Gryder, Bruce Hall, Walt Tankersly, Bill Simmons, Bill Bodie, Bruce Lytle, Charles Peigler, and William Nelms. These people were either civil servants or contractors that worked at Huntsville or at the Cape.

George Moore, 1958 Clemson graduate also worked in the ABMA Missile Firing Lab at Cape Canaveral in the early days of the US Army Missiles Development Program. He and his team worked on down range missile impact points and trajectories. He was in the blockhouse when the ABMA team launched the Explorer IV satellite and for the early test launches of the Jupiter IBM series missiles.

Col William McPherson Clemson Class of 30 was the Commandant of the Guided Missile School here at Redstone Arsenal during the development of the Army's Jupiter Missile Program.

I am sure there were many other Clemson graduates that were employed at other NASA Centers and or Contractors supporting the Missile and Space Programs but they were not brought to my attention during my research for this book.

BIBLIOGRAPHY

1. The Tiger Newspapers
2. The TAPS Year Books
3. The Anderson Independent and Daily Mail Newspapers, Anderson, SC
4. The Greenville Daily Newspaper, Greenville, SC
5. The Put-Put Air Force, The Story of the Civilian Pilot Training Program and The War Training Service(1939-1944), Department of Transportation Federal Aviation Adminstration, Aviation Education Staff, GA-20-80, Strickland, Patricia, 1971.
6. To Fill the Skies with Pilots, The Civilian Pilot Training Program, (1939-1949), Pisano, Dominick, Urbana: University of Illinoise Press, 1993.
7. Wheels Up, The Adventures of a Private Pilot, Nesbit, James D., New River Press, Woonsocket R.I. 1988
8. Clemson University College of Engineering, One Hunderd Years of Progress, Benjamin, Laura L. Clemson University, 1989
9. Personal Communication, Morgan, Herbert
10. Personal Communication, Harper, John E.
11. Personal Communication, Wigington, Max
12. Personal Communication, Anthony L. Meldeau
13. Scholars in Foxholes, The Story of the Army Specialized Training Program in World War II, Keefer, Louis E., COTU Publishing, Reston, Virginia, 1988.
14. American Aviation Historical Society
15. Ninety-Nines Inc, International Organization for Women Pilots.
16. USAF Biographies via Internet
17. Air Force Historical Research Agency, Maxwell AFB, Alabama
18. The Pre-Astronauts, Manned Ballooning on the Threshold of Space, Craig Ryan, Naval Institute Press, Anapolis, Md, 1995.
19. Man High, Lt Col David G. Simons and Don A. Schanche, Doubleday and Company Inc, Garden City, NY, 1960
20. Personal Correspondence Jack Nesbit, Jim Nesbit's son
21. Autobiography of Values, Charles A. Lindbergh, A Harvest/HBJ Book, Harcourt Brace Jovanich, Publishers, San Diego, New York, London.
22. On Celestial Wings, Col Edward Whitcomb, Chapter 21,Boselli and the Sacred Cow, Air University Pres, Nov 1995.
23. Air Pilots Register, 1935, Air Pilots Registery Co., 45th West 45 Street, New York, New York
24. Military Science (Army Rotc and Aerospace Studies) - AFROTC - Information Access Policy, Clemson Universities, B. Ellis, summer 2000

Some Sources for Clemson College's History during WWII

McKale, Donald M. Destined for Duty: The Clemson Class of 1941 Clemson, SC: Clemson University, 1990.

Benjamin, Laura L. Clemson University College of Engineering One Hundred Years of Progress Clemson University, 1989.

Howard, Frank and Friends. Howard 1990.

The Tiger, 1941-1946

Taps, 1941-1943, 1946

Annual Reports of the Board of Trustees of Clemson Agricultural College to the General Assembly of South Carolina, 1942-1946

John D. Lane Papers. Clemson University Libraries Special Collections.

Sydney Crouch Papers. Clemson University Libraries Special Collections.

R. F. Poole Papers. Clemson University Libraries Special Collections.

Robert M. Jones Papers

Janet Moja and Rose Davis, April 3, 1991

Clemson Agricultural College Catalogue, 1942-1946

Sorrells, Robert. The Clemson Experimental Forest: Its First Fifty Years 1984.

Minutes of the Meetings of the Deans and Directors of Clemson Agricultural College, 1941-46. Clemson University Libraries Special Collections.

Minutes of the General Faculty of Clemson Agricultural College, 1940-1947. Clemson University Libraries Special Collections.

Annual Reports of the Registrar's Office, 1941-1946. Clemson University Libraries Special Collections

Presidents' Reports to the Clemson College Board of Trustees, 1941-1948, Clemson University Libraries Special Collections

Daily Bulletins, Army Service Forces, Clemson Agricultural College, Clemson, SC, 1944-1945, Clemson University Libraries, Special Collections

"Tips on the Tigers," 1946-47

Susan Hiot, Research Summary for Exibit "World War II and the Clemson Community," Dec 2, 1991 through June 1992

Information from: LaVern Pyles, Jr., USN (Ret.), Class of '43
Dr. William H. Hunter, MD
S. F. Phillips, US Army (Ret.), Class of '42

The National Encylopedia of Biography(New York: James T. White and Company, 1935 vol XXIV.

"American Air Service Units with the British Expeditionary Forces" Gorrell History, Air Service, American Expeditionary Force, series b, vol XII.

Hostile Skies: The Combat History of the American Air Service in World War I, James J.Hudson, Syracuse University Press, 1968.